Matthias Busker

IR-UV Spectroscopy of Chromophores with π-Systems

Matthias Busker

IR-UV Spectroscopy of Chromophores with π-Systems

and their Clusters in the Gas Phase

Südwestdeutscher Verlag für Hochschulschriften

Impressum / Imprint
Bibliografische Information der Deutschen Nationalbibliothek: Die Deutsche Nationalbibliothek verzeichnet diese Publikation in der Deutschen Nationalbibliografie; detaillierte bibliografische Daten sind im Internet über http://dnb.d-nb.de abrufbar.

Bibliographic information published by the Deutsche Nationalbibliothek: The Deutsche Nationalbibliothek lists this publication in the Deutsche Nationalbibliografie; detailed bibliographic data are available in the Internet at http://dnb.d-nb.de.

Verlag / Publisher:
Südwestdeutscher Verlag für Hochschulschriften
ist ein Imprint der / is a trademark of
OmniScriptum GmbH & Co. KG
Heinrich-Böcking-Str. 6-8, 66121 Saarbrücken, Deutschland / Germany
Email: info@svh-verlag.de

Herstellung: siehe letzte Seite /
Printed at: see last page
ISBN: 978-3-8381-1153-7

Zugl. / Approved by: Düsseldorf, Heinrich Heine Universität Düsseldorf, Diss., 2009

Contents

1 Introduction

Biomolecules, as DNA-bases and proteins played a key role in the development of life forms as we know them nowadays. The first life forms had to create photostability mechanisms, which still protect us from UV photo damage. These mechanisms allow the photo excited molecules to relax to the electronic ground state on a fast time scale, which is often done by passing conical intersections (CI's). These represent intersections of two states in a three dimensional hyper potential surface and often allow barrierless transition from the excited state to an intermediate state or the hot electronic ground state.

1-Methylthymine is a model system for the DNA-base thymine, since thymine is covalently bound to the phosphate backbone in the 1 position in DNA double helices. Excited 1-methylthymine molecules show fast transitions to a dark intermediate state, in which the excited molecules stay for approximately 230 ns (in the gas phase). This state might play a role as precursor state for the cyclobutane pyrimidine dimer (CPD) formation, which is suspected to create skin cancer.

A complete understanding of the pathways and states involved in 1-methylthymine has fascinated scientist for more than 50 years. Our contribution to this field is the possibility to investigate the properties of 1-methylthymine and it's clusters with water under isolated conditions (without effects created by solvent or matrix molecules). The expansion of the molecules in supersonic gas jets and the spectroscopic analysis with resonance enhanced multi photon ionization and IR/UV double resonance spectroscopy in combination with time of flight mass spectrometry allows to characterize the molecule's and cluster's structures and lifetimes.

Besides the intrinsic intramolecular features enabling photo stability, the structure of a biomolecule influences it's photochemical properties as well. Intermolecular interactions, which often are hydrogen bonds or van der Waals bonds define the molecular structure. These interactions give a large structural flexibility to the molecules due to binding energies of less than 40 $kJmol^{-1}$. The strong flexibility further influences the photostability as well, since relaxation pathways via conical intersections often require significant structural changes.

Model systems showing dispersive interactions are of great interest, since CH...π, CH...O and CH...N interactions dominate the structure of proteins and have neither pure van der Waals nor hydrogen bond character. An investigation of these interactions in model systems can give a deeper understanding of the structure defining parameters in proteins, for example. Furthermore, dispersive interactions are responsible for the structure of crystals, as these attractive forces control the later structure of nano crystallites by which a co-crystallization process is started (seed crystal).

Laser spectroscopy in molecular beams in combination with time of flight mass spectrometry is used in this thesis to unravel pathways in crystal formation processes. Starting from single molecules, the formation of clusters consisting of up to four molecules is observed. These structures are possible precursors for later crystals (nano crystallites, seed crystals). IR/UV double resonance spectroscopy in molecular beams combined with time of flight mass spectrometry enables the experimentalist to study the stepwise kinetically controlled growth of nano-crystallites in an mass and isomer selective manner.

2 Theoretical Background

In order to characterize the aggregation, isomerisation and conformation of molecules and of the electronic states involved in those processes, one has to use several experimental and theoretical approaches.

Time of flight mass spectrometry in combination with supersonic expansion is a suitable tool to significantly reduce the population of higher molecular vibrational, translational and rotational states. By this, the number and width of spectral features is reduced significantly (see section 2.1).

The skimmed molecular beam is then analyzed by different laser spectroscopical techniques including resonance enhanced multi photon ionization spectroscopy (REMPI, see section 2.2.1), IR/UV double resonance spectroscopy (see section 2.2.3) and delayed ionization spectroscopy (see section 2.2.2). The interpretation of recorded spectra is done by comparison to quantum chemical calculations (see section 2.3). These often reflect the experimental results well and give direct access to specific vibrations in infrared spectroscopy, for example.

Furthermore, the interpretation of cold (temperatures in the order of 10 Kelvin) [1] gas phase spectra is much easier than in the liquid phase, since effects by molecular collisions, hot bands or matrix effects can be ruled out.

The idea of this chapter is to give a brief insight to the methods used in this thesis.

2.1 Molecular Beams

2.1.1 Isentropic Expansions in Circular Orifices

The carrier gas, which contains the sample molecules is expanded from a sample oven at pressures of some hundred mbar up to 4 bar through a small pinhole (300 to 500 μm) into vacuum (10^{-6} mbar) and forms a free jet.

In order to perform spectroscopy in a cold region of the jet, one has to be aware of the temporal profile and the spatial configuration of the free jet. Of spectroscopic interest is only the zone of silence [1], which is suitable for laser spectroscopy in combination with time of flight mass spectrometry. Different types of shock s emerge in the boundary of the jet, in which the molecular temperature and speed decrease drastically. A schematic of a typical isentropic expansion through a pinhole is presented in figure 2.1.

The molecules in the free gas jet are continously accelerated during expansion. Since the molecules are travelling at supersonic speed, the gas density decreases at increasing distance from the nozzle until the gas pressure is well below the residual pressure inside the vacuum chamber (p_r). The over-expanded gas jet by this creates a shock front under recompression (Barrel shock in figure 2.1).

Another shock front formed is the so-called "'Mach's disc"'. This shock front emerges, as the speed at which the gas jet is travelling decreases again and reaches the sonic speed of the carrier gas.

The position of the Mach disc is defined by equation 2.1 [1].

$$\frac{x}{d} = \sqrt{0.67 \cdot \frac{p_0}{p_r}} \tag{2.1}$$

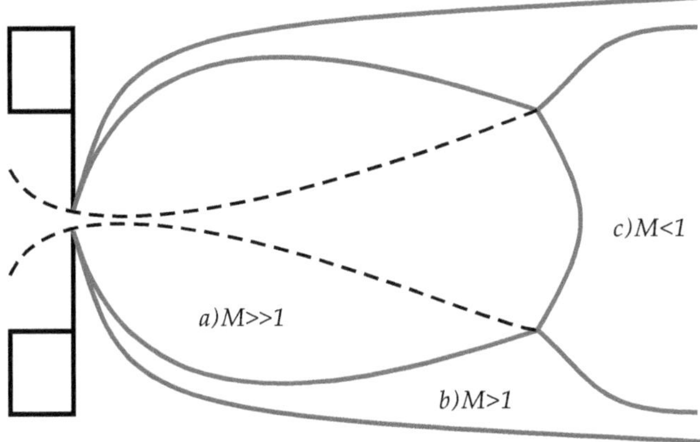

Figure 2.1: *Gas flow structure in isentropic expansions [1]*

Where x is the distance of the Mach disc to the nozzle, d is the orifice diameter, p_0 is the reservoir pressure and p_r is the residual pressure in the vacuum chamber. Hence, the Mach disc position only depends on the expansion pressure and the pump ratio of the vacuum pumps and not the type of noble gas used. One has to know the exact position of the Mach's disc during pulsed expansion, which changes temporally due to the changing p_r. The Mach's disc has to be situated behind the skimmer during the whole expansion process, since the shock waves reflected by the skimmer would otherwise impede the formation of a molecular beam.

The Mach number inside the Barrel shock is slightly above 1. This rather hot region is directly removed by the skimmer in the later molecular beam setup (discussed in chapter 3.2) and only the cold zone of silence ($M >> 1$) is used for spectroscopy. Hence, the valve timing has to be chosen such that only the zone of silence (a quasi-plateau) is detected and neither the Mach disc nor slow and hot residual molecules late after the zone of silence are detected. It has proven to be useful to perform spectroscopy 50 to 150 μs after the very beginning of the gas pulse (for pulse lengths of 250 to 400 μs).

The term zone of silence (also called Mach's bottle) results from the fact that the very low temperatures and pressures in this zone lead to a very low speed of sound and thus very high reachable Mach numbers (Mach numbers of 20 can be reached locally).

2.1.2 Molecular Cooling

The energy accumulated in the molecule's energetic degrees of freedom is transferred by collisions in the nozzle to the carrier gas molecules. Vibration, rotation and translation are cooled separately and with different efficiencies (kinetically controlled) [2].

One has to differentiate two cases, which are defined by the Knudsen Number presented in equation 2.2.

$$K_{\mathrm{n},0} = \frac{\lambda_0}{d} \tag{2.2}$$

Where $K_{n,0}$ is the Knudsen number, λ_0 is the molecule's mean free path and d is the nozzle diameter [1].

If $K_{n,0}$ is much larger than 1, the molecules hardly undergo any collisions such that there is no cooling effect and the molecular velocity distribution remains unchanged.

If $K_{n,0}$ is lower than 1, the number of molecular collisions is in the order of 10^2 and active cooling occurs.

The translational energy is equalized for all molecules in the expanding gas jet. This is the main reason, why noble gases are used as carrier gases, since they are inert and can be cooled quite efficiently. The translational energy parallel to the jet's axis is described by a narrow Dirac distribution [2]. The translational energy distribution perpendicular to the jet is effectively reduced by skimming the beam.

The rotational temperature is cooled quite efficiently by collisions, such that 10^2 collisions are sufficient [1] to significantly reduce rotational energy. The vibrational temperature is reduced less efficiently, such that up to 10^4 collisions are needed [1]. Since there are hardly any collisions in the later molecular beam (skimmed free jet) due to the narrowed speed distribution, 10^4 collisions are often not reached in isentropic expansions and thus the vibrational temperature remains the highest (2.3).

$$T_{\text{translational, }\perp} < T_{\text{translational, }||} < T_{\text{rotational}} < T_{\text{vibrational}} \tag{2.3}$$

Where $T_{translational,\ \perp}$ is the molecule's translational temperature perpendicular and $T_{translational,\ ||}$ parallel to the beams direction of travel. Temperatures of a few K, which directly reflect the molecule's rotational temperature are reached by this method.

2.1.3 Maximum Particle Velocities

From an experimental point of view, it is very important to know the approximate velocity, at which the gas molecules are traveling along the vacuum chamber after expansion to set up proper timing conditions in these pulsed experiments. The maximum jet velocity u can be calculated, when considering energy conservation given by equation 2.4.

$$h + \frac{1}{2} \cdot u^2 = constant \tag{2.4}$$

Where h is the mass dependent enthalpy of an ideal gas. Basically, this equation just describes the sum of the potential energy (heat energy of the gas) and the kinetic energy of the flowing gas. The enthalpy h can now be expressed as a function of the temperature independent heat capacity c_p and the temperature T.

$$c_{\mathrm{p}} = \frac{h}{T} \text{ or } h = c_{\mathrm{p}} \cdot T \tag{2.5}$$

By inserting equation 2.5 in equation 2.4, we get the expression:

$$c_{\mathrm{p}} \cdot T + \frac{1}{2} \cdot u^2 = constant \tag{2.6}$$

Since the system's energy is conserved before and after expansion and as the mean velocity of the gas molecules prior to expansion can be considered to be very small compared to the mean velocity during expansion, equation 2.7 can be set up. T_0 and T are the gas temperatures in the reservoir and during expansion, respectively.

$$c_{\mathrm{p}} \cdot T + \frac{1}{2} \cdot u^2 = c_{\mathrm{p}} \cdot T_0 \tag{2.7}$$

When considering the specific heat capacity c_p as temperature independent in this case, it can be expressed as:

$$c_{\mathrm{p}} = c_{\mathrm{v}} + R_{\mathrm{s}} \tag{2.8}$$

Since c_p depends on c_v, an expression for the volume dependent specific heat capacity has to be introduced, which is given by:

$$c_{\mathrm{v}} = \frac{f}{2} \cdot R_{\mathrm{s}} \tag{2.9}$$

By inserting equation 2.9 in equation 2.8, the final term for c_p can be defined as:

$$c_{\mathrm{p}} = R_{\mathrm{s}} \cdot (\frac{f}{2} + 1) \tag{2.10}$$

f represents the number of the molecule's energetic degrees of freedom and is ≥ 3. In case of noble gases, f has to be 3, since only 3 translational, 0 rotational and 0 vibrational degrees of freedom are available.

R_s is the specific gas constant, which is defined as:

$$R_s = \frac{R}{M} \tag{2.11}$$

where M is the molar mass and R is the ideal gas constant. By this, one can calculate typical specific gas constants and pressure dependent heat capacities for the carrier gases helium, argon, neon and xenon using the ideal gas constant R = 8,3145 $\frac{J}{Mol \cdot K}$ (presented in table 2.1).

Noble gas	Molar mass [g·Mol^{-1}]	R_s [J· g^{-1}K $^{-1}$]	c_p [J· g^{-1}K $^{-1}$]
Helium	4.00	2.079	5.198
Neon	20.18	0.403	1.008
Argon	30.95	0.269	0.673
Xenon	131.29	0.063	0.158

Table 2.1: *Typical R_s and c_p values for some noble gases*

Rearranging equation 2.10 gives the maximum velocity u. The expression is presented in equation 2.12.

$$u = \sqrt{2 \cdot c_p \cdot (T_0 - T)} \tag{2.12}$$

Typical calculated maximum expansion velocities can be found in table 2.2.

The maximum jet velocity is an important feature defining the trigger delay for the pulsed valve, which has to be chosen such that the pump and probe lasers, the ion optics and the data acquisition analyze a cold section of the molecular beam.

One can calculate the gas jet travelling times, when considering the distance between the orifice and the place of ionization to be 180 mm. Using the calculated velocities, one can estimate that the molecular beam needs the time t presented in table 2.2 to travel from the nozzle to the place of ionization.

Noble gas	u [m· s^{-1}]	t [μs]
Helium	1685	1.068
Neon	742	2.426
Argon	606	2.970
Xenon	294	6.122

Table 2.2: *Maximum expansion velocities u and travel times for some noble gases at 293K (Room temperature, T_0) and 20K (T)*

2.2 Spectroscopical Methods

The spectroscopical methods used in this thesis are resonance enhanced multi photon ionization (REMPI), delayed ionization and IR/UV double resonance spectroscopy. The first two methods are used to characterize the electronic structure of molecules and IR/UV spectroscopy applied to molecular beams and in combination with time of flight mass spectrometry is a very valuable tool for recording mass, isomer and conformer selective infrared spectra. These techniques allow structural assignment and cluster identification by comparison to quantum chemical calculations.

The basic principles and ideas of the spectroscopic methods used in this thesis are going to be described in the next sections.

2.2.1 Resonance Enhanced Multi Photon Ionization

Resonance enhanced multi photon ionization (REMPI) applied on cold molecules in combination with time of flight mass spectrometry allows the investigation of molecules under very controlled conditions. By absorbing two or more photons, the molecule is first excited and then ionized (pump and probe experiment).

One has to differentiate between one and two color REMPI in which photons of equal energy (1+1, one color) or photons of differing energy (1+1', two color) are absorbed. The method mostly used in this thesis is 1+1' R2PI (two color resonant two photon ionization), in which two photons of different wavelengths are absorbed to reach the ionization potential (IP).

The molecule under investigation absorbs a single UV photon (excitation laser) and gets excited to a higher singlet state having a certain intrinsic lifetime. The second photon (ionization laser) is absorbed instantaneously and overcomes the first ionization energy turning the molecule into a cation.

The photon energies are chosen such that the molecule undergoes a resonant transition to an excited singlet state and gets ionized by a second photon of the same or different wavelength. The efficiency of this resonant step is much higher than for a non-resonant two photon step. The principle energetic transitions in REMPI spectroscopy are presented in figure 2.2 section 1.

The ionized molecules pass a linear time of flight mass spectrometer (Wiley McLaren configuration) and are detected by multi channel plates (MCP).

One color REMPI (figure 2.2 section 1a) is a very simple tool to characterize the vibronic properties of molecules, since two photons of the same laser pulse can be absorbed, which significantly simplifies the experimental setup. For fragile clusters or molecules, whose excited singlet state is situated below half of the ionization energy two color REMPI is favorable.

If the molecule is excited by a photon having the energy $E=h\cdot\nu_1$ and a higher energy photon ($E=h\cdot\nu_2$) is needed for the ionization step, only two color REMPI allows to record spectra. By this, molecules not suitable for one color REMPI due to their high ionization potentials can be characterized. Furthermore, molecules undergoing a transition to long lived intermediate states after excitation can be ionized from those states using a short ionization wavelength of 193 nm, for example. This often increases the signal to noise ratio significantly.

In the case of clusters or molecules having an excitation energy above half the ionization energy, the pump photon energy has to be high to reach the IP in a one color process. By this a large amount of excess energy is conserved in the ions, which is transferred to the molecule's vibrational degrees of freedom and leads to fortified fragmentation to the lower lying cluster and monomer masses. Due to this effect the spectra of dissociating clusters can sum up on all fragment masses. A way to

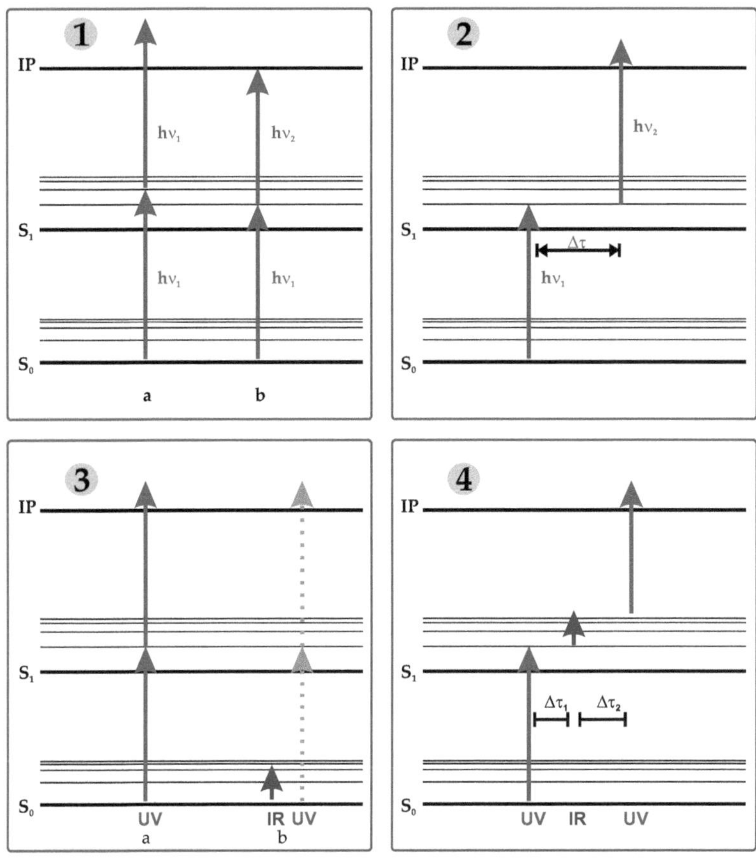

Figure 2.2: *Schematic showing* ***1.*** *1+1 (a) and 1+1' (b) REMPI spectroscopy* ***2.*** *delayed ionization spectroscopy* ***3.*** *IR/UV double resonance spectroscopy in the ground state(SHB) and* ***4.*** *IR/UV double resonance spectroscopy in the electronically excited state*

reduce fragmentation is careful adaption of excitation and ionization energies using two color REMPI spectroscopy.

2.2.2 Delayed Ionization

The photostability of biological molecules is often based on ultra fast decay mechanisms via conical intersections (CI), which lead to the molecules hot ground state. In case of 1-methylthymine, the molecule is excited to the S_2 ($^1\pi\pi^*$) state and a fraction of the molecules undergo a fast transition to the S_1 ($^1n\pi^*$) state (see section 4). This spectroscopically dark intermediate state can be ionized after a certain delay time. By this, direct measurent of the dark state's lifetime is possible by varying the laser delay time and measuring the delay dependent ion signal intensity

Furthermore, delayed ionization offers a new range of techniques for IR/UV double resonance spectroscopy, as discussed in the following section. A schematic of a delayed ionization pump probe experiment is presented in figure 2.2 section 2.

2.2.3 IR/UV Double Resonance Spectroscopy

The basic principles of IR/ UV double resonance spectroscopy or spectral hole burning are presented in figure 2.2 section 3 (also called RIDIR (resonant ion depletion infrared spectroscopy) in literature).

The REMPI process as presented in section 2.2.1 strongly depends on the population of the ground and excited electronic states. The photon wavelengths are chosen such that the vibronic ground state is pumped and the first excited electronic state is probed. When irradiating the molecule with an infrared (IR) photon some hundred nano seconds priorly to the excitation laser, a depopulation of the cold electronic ground state occurs by vibrational excitation (see figure 2.2 section 3b). Due to the reduced number of molecules in the vibronic ground state, the resonant ion signal produced by the REMPI process decreases.

If one now fixes the excitation and ionization laser to resonant wavelengths and scans the IR laser, a decrease of the ion signal at distinct IR wavelengths, which correspond to certain vibrational transitions from the cold electronic ground state can be observed as dips in the ion signal. This technique in combination with molecular beams and time of flight mass spectrometry allows to record isomer and mass selective ground state IR spectra, if the investigated isomers have well separated UV resonances.

The IR spectra often easily unveil structural properties of the monomers and clusters, when compared to adapted calculations. Furthermore, clusters that tend to fragment and create ion signals on lower lying masses can be differentiated by comparison to calculated IR spectra.

In another technique, the two UV lasers are delayed by for example 50ns and the IR laser is temporally placed in between them. This type of spectroscopy allows to probe the electronically excited state. See figure 2.2 section 4.

When considering fragile clusters for example, a resonant vibrational excitation can lead to controlled fragmentation and thus to an increase in monomer ion signal. IR spectra of clusters can be recorded on the monomer mass in this way.

2.3 Quantum Chemical Calculations

Quantum chemical calculations are a very important tool when it comes to interpret IR and UV spectra. Using calculations adapted to the system under investigation allows to assign structures to recorded spectra, which would otherwise require a larger experimental effort.

In this thesis, a large number of different cluster structures and sizes are discussed. In case of benzene (B) acetylene (A) clusters for example (see sections 8 and 9) 27 different clusters are discussed and according structures assigned to the IR spectra of $B_1A_{1\text{-}3}$ and $B_{1\text{-}2}A_{1\text{-}2}$.

Most of the calculations were performed by Thomas Häber and Mihajlo Etinski using the TURBOMOLE and MOLPRO program packages [30, 34]. Furthermore, the GRANADA program [31] which randomly arranges molecules and by this creates several hundred possible cluster structures in combination with the MOPAC2007 program package [33] which performs pre-optimizations on the PM6 level [32] allows a comprehensive and impartial analysis of possible cluster structures.

Since the investigated systems are significantly controlled by $CH \cdots N$, $\pi \cdots \pi$ or $CH \cdots \pi$ interactions or hydrogen bonds, theoretical methods incorporating dispersion interactions have to be preferred to best describe the situation. Hence, recent calculations were performed on the DFT-D level of theory.

3 Experimental Setup

Many different electrical, mechanical and optical devices have to be combined to form the setup used in this thesis. Since the experiments take place in a high vacuum, turbo molecular pumps and oil diffusion pumps are combined with roots and rotary vane pumps to reach end pressures of 10^{-6} to 10^{-7} mbar.

The electro-optical setup consists of two UV dye lasers, an IR OPO/OPA laser system and an Ar/F_2 excimer laser. The created cations are deflected and analyzed by a linear time of flight mass spectrometer in Wiley McLaren configuration and detected by a multi channel plate (MCP) detector. Data acquisition is done using a 500 MHz digital oscilloscope and a computer system with dedicated software written in National Instruments LabView. A schematical representation of the basic experimental setup and the control system is presented in figure 3.1.

The scope of this section is to describe the basic elements of the lasers, the molecular beam source and the time of flight mass spectrometer.

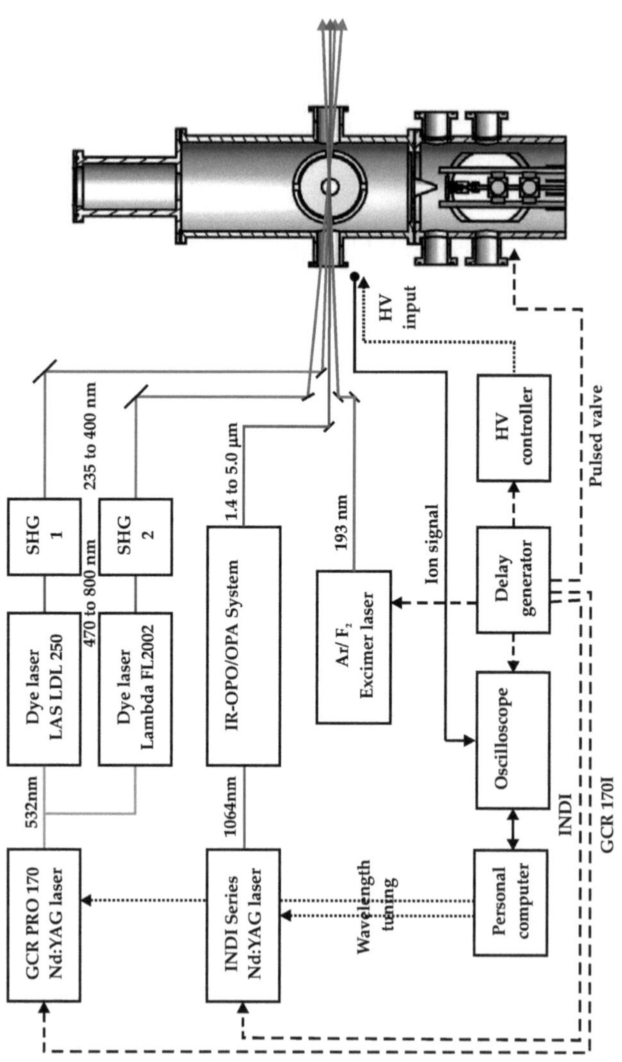

Figure 3.1: *Schematic overview of the IR/UV double resonance setup incorporating all possible laser configurations, data acquisition and timing elements*

3.1 Laser Systems

3.1.1 The UV Lasers

The UV lasers used in this setup are two different dye lasers (LAS LDL 250 and Lambda FL2002) and an Ar/F_2 excimer laser.

The dye lasers (Lambda FL2002 and LAS LDL 205) are pumped by a frequency converted Spectra Physics GCR Pro 170 Nd:YAG laser (532 nm or 355 nm depending on the pumped laser dye) and the visible laser output is frequency doubled by a non linear optical crystal (KDP and BBO, depending on the fundamental wavelength). The principles of Nd:YAG and dye lasers [3, 4, 5, 6] and non-linear optical crystals [6, 7, 5] are well understood and described elsewhere. The LAS LDL 250 dye laser can be continously scanned over a large range of UV and visible wavelengths having a linewidth of approximately 0.2 cm^{-1}. It was equipped with new stepper motors by Thomas Häber and calibrated using a Neon hollow cathode tube to allow direct wavelength tuning by dedicated LabView software. The Lambda FL2002 laser can only be operated at a fixed wavelength at a linewidth of approximately 0.2 cm^{-1}.

The argon fluoride excimer laser is a gas laser system emitting at a fixed wavelength of 193 nm. The basic principles of excimer laser operation can be found in [3]. The short wavelength of the excimer laser, which can hardly be produced with dye lasers or UV OPOs is ideal to record time of flight mass spectra of all kinds of molecules and ionize molecules having a large ionization potential.

3.1.2 The IR OPO/OPA System

The IR OPO/OPA (Infrared optical parametric oscillator/ optical parametric amplifier) creates laser radiation in the wavelength regime of 1.35 to 5 μm at a linewidth of approximately 0.05 cm^{-1} when pumped with a seeded Nd:YAG laser. Most of the experiments were carried out using the IR-OPO system pumped by a unseeded Nd:YAG laser. In this configuration the infrared bandwidth increases to $1cm^{-1}$. The basic setup is shown in figure 3.2.

The 1064 nm pump radiation coming from a Spectra Physics Indi laser enters the IR OPO/OPA via the entrance pupil (1). After reducing the beam's diameter by a telescope (2), the pump laser is split up into two separate paths by a beam splitter (3). Approximately 40% of the laser light is directed into the OPO path through a Type 2 frequency doubling crystal (KTP, 4), which converts the IR light to visible radiation at 532 nm. The green light is directed via the mirrors (5, 6, 7) onto the holographic grating (8), which reflects the light (0th order of diffraction) through the OPO crystal (KTP, 9). The incident 532 nm light is transferred into infrared signal and idler waves ($\lambda_{\text{SIGNAL}} = 710 - 880nm < \lambda_{\text{IDLER}} = 1346 - 2124nm$) via optic paramatric oscillation (see [6, 8]) and reflected by the end mirror (10). After passing the OPO crystal a second time, the IR light is reflected by the holographic grating. A small portion of the IR light is diffracted into the grating's -1st order and directed towards the mirror (11). The reflected light is then partially guided back into the OPO crystal by diffraction to the grating's 1st order. By this, a resonator structure incorporating a wavelength selective element (grating and mirror (8, 10, 11)) is created, which seeds the emission wavelength for the DFM (difference frequency mixing, see [6, 7]) process in the KTP crystal (9). The signal and idler beam's polarization is rotated by the Berek compensator (12, a variable $\lambda/2$ retardation plate, [8]) and the beam's dimensions are corrected by a tilted dove prism (13) to fit a square-like shape.

The remaining pump radiation (60%) is led by two mirrors (14, 15) through the four OPA crystals (KTA, 16, Optical Parametric Amplifier). By overlapping the idler wavelength from the OPO path and the 1064 nm pump beam in the OPA crystals, a

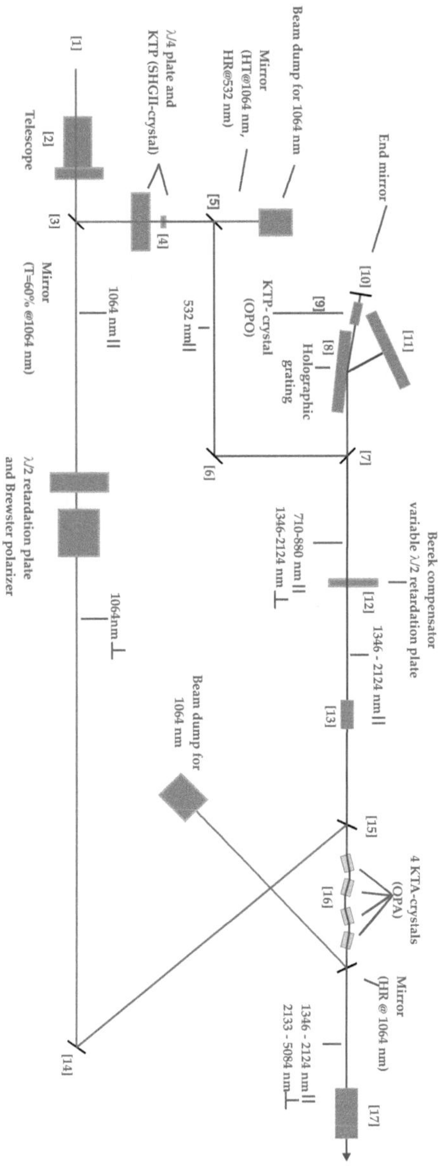

Figure 3.2: *Schematic showing the opto mechanical setup of the new OPO system*

new pair of signal and idler waves is created by difference frequency mixing (λ_{SIGNAL} = 1346 - 2124 nm and λ_{IDLER} = 2133 - 5084 nm). The signal wave coming from the OPO path is s-polarized and passes the OPA crystals unaltered. Since the new signal and idler waves are perpendicularly polarized, they can be separated using a brewster polarizer (17).

The OPO path's p-polarized idler wave in this setup directly defines the wavelength of the created and amplified signal and idler waves in the OPA path. Hence, the system's output wavelength can be scanned by tuning the OPO crystal (9).

This setup enables continous wavelength scanning from 1346 to 2124 nm and 2133 to 5084 nm with a gap of only 9 nm at a bandwidth of approximately 1cm^{-1} (0.05 cm^{-1}, when operated with a seeded Nd:YAG laser in the future). The only operation to be done when scanning the gap is to rotate the Brewster polarizer to choose between signal and idler waves.

3.2 Linear Time of Flight Mass Spectrometer

The time of flight mass spectrometer consists of a source chamber with a dedicated pulsed free jet molecular source and a skimmer, an Ionization chamber with repeller and deflection plates and a flight path with a multi channel plate detector (MCP). A schematical representation of the linear time of flight mass spectrometer (TOF) is presented in figure 3.3.

The substance under investigation is heated (Maximum temperature of 180°C, limited by the pulsed valve) and partially evaporated in the sample oven. The molecules get picked up by a carrier gas (helium or argon) and adiabatically expanded into the high vacuum ($2 \cdot 10^{-6}$ mbar) through the pulsed valve (stagnation pressures of some hundred millibars to 4 bar against vacuum and diameters of 500 or 300 μm). The source chamber is evacuated by an oil diffusion pump (Pfeiffer Dif 250, 1800 Ls^{-1}).

The molecules cool down during expansion, as described in section 2.1 and the cold free jet is clipped by the skimmer after a travel of approximately 2.5 cm (Beam dynamics, diameters of 0.7 to 2 mm). The described situation is presented in figure 3.4. The pressure inside the source chamber increases to values of some 10^{-4} mbar during expansion. A detailed presentation of the source in combination with a skimmer is shown in figure 3.4.

The pulsed molecular beam then crosses the perpendicularly traveling laser pulses and the molecules are excited and ionized in one of the ways described in section 2.2. The pressure inside the ionization chamber is kept at some 10^{-6} mbar without molecular beam and increases to 2 - $5 \cdot 10^{-5}$ mbar during expansion to maintain a differential vacuum. The cations are rejected by the repeller plates and accelerated by the deflection plates. Since a voltage ramp is applied to the deflection plates,

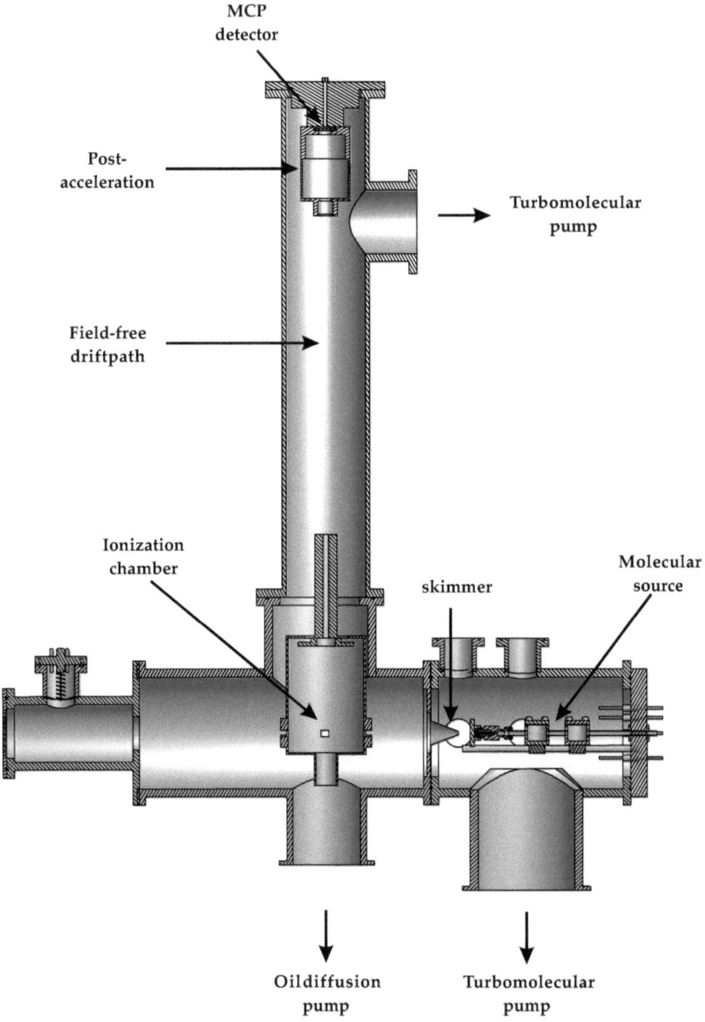

Figure 3.3: *Schematic overview of the time of flight mass spectrometer*

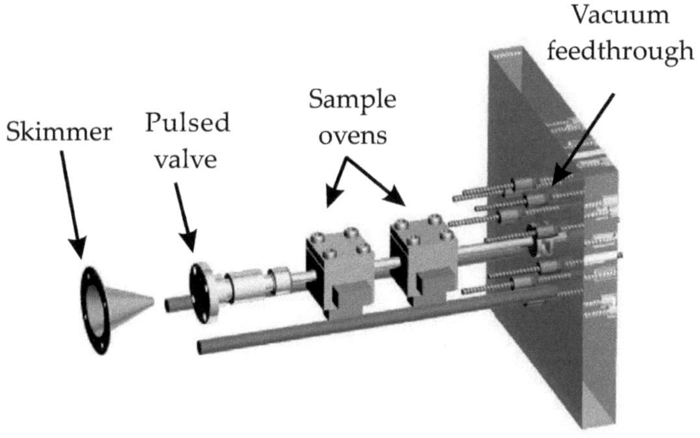

Figure 3.4: *Detailed view on the molecular source*

background ions, which do not originate from the molecular beam can be reduced via the ion stop. By this, a controlled steering of the ion beam onto the MCPs is possible.

The ions drift along the field-free drift path and are again accelerated by the post accelerator plates, which allow a higher detection yield of molecules hitting the MCPs. This setup is especially helpful for heavier molecules, which would be hardly detectable due to their rather low velocities. The pressures inside the drift path are in the order of 10^{-7} mbar. The situation is presented in figure 3.5.

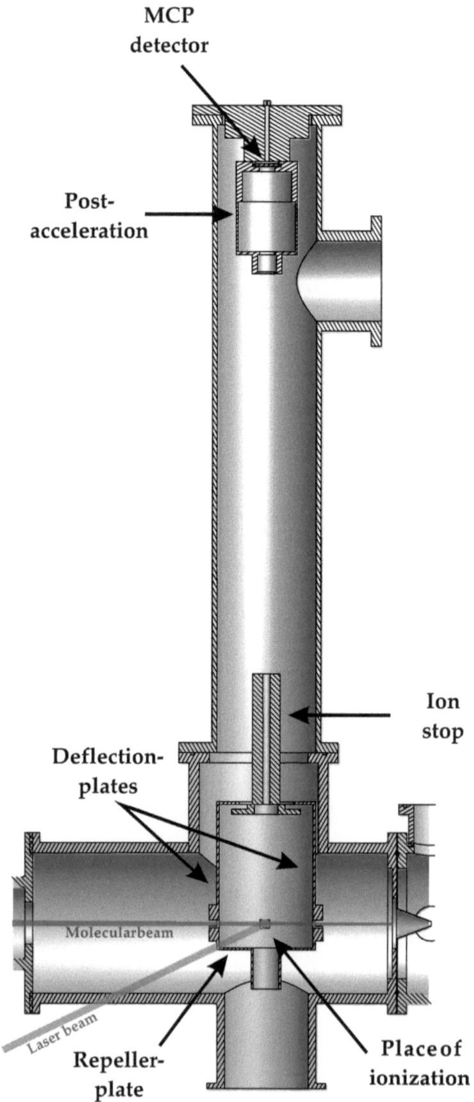

Figure 3.5: *Schematical representation of the ionization chamber, drift path and detector.*

3.3 Temporal Management

The lasers are operated at repetition rates of 10 Hz for the Spectra Physics Indi and excimer laser and 20 Hz for the GCR Pro 170. Hence, the data acquisition in this experiment is locked to 10 Hz. In order to time the single stages of the experiment, which are the pulsed valve, the lasers, the voltage ramp/ MCP detector and the oscilloscope, trigger signals coming from a main trigger are delayed by three different delay generators. One off the shelf Stanford research systems (DG535) and two house-build delay generators. Since the experiment strongly depends on proper timing, typical delays for helium and argon expansion are stated in tables 3.1 and 3.2.

Delay generator	Channel	Delayed component	Timing T_0 + x [ms]
Stanford	A	Voltage ramp	2.09444
Stanford	B	Oscilloscope	2.09719
Stanford	C	Pulsed valve	0.00050
Clock 1	T2	Q-switch GCR-170	2.09380
Clock 2	T4	Flash lamps GCR-170	1.92760

Table 3.1: *Typical timings when expanding in helium*

Delay generator	Channel	Delayed component	Timing $T_0 + x$ [ms]
Stanford	A	Voltage ramp	2.09584
Stanford	B	Oscilloscope	2.09719
Stanford	C	Pulsed valve	0.00031
Clock 1	T2	Q-switch GCR-170	2.09380
Clock 2	T4	Flash lamps GCR-170	1.92760

Table 3.2: *Typical timings when expanding in argon*

3.4 Data Acquisition and Management

The signal detected by the MCPs and recorded by the digital oscilloscope (Tektronix TDS 520A) is transferred via a GPIB Bus (IEEE 488.2) to a Personal Computer (PC). The PC records the oscilloscope signals and controls the laser wavelengths and oscilloscope settings via a LabView program (National Instruments). The control scheme is presented briefly in the overview in figure 3.1.

The software enables the user to choose between the separate tuning elements of the lasers, change the oscilloscope settings and record spectra in a wide time gate. By this, larger TOF spectra correlated to the laser wavelengths and thus two dimensional spectra can be recorded. This is very helpful when simultaneously recording spectra of several mass traces (for example different cluster sizes or species).

4 1-Methylthymine and it's Water Clusters

Electronic and Vibrational Spectroscopy of 1-Methylthymine and its Water Clusters: The Dark State Survives Hydration

M. Busker[1], M. Nispel[1], Th. Häber[1], K. Kleinermanns[1*],
M. Etinski[2], T. Fleig[2*]

[1] Institut für Physikalische Chemie und Elektrochemie 1, Heinrich-Heine Universität-Düsseldorf, 40225 Düsseldorf, Germany

[2] Institut für Theoretische und ComputerChemie, Heinrich-Heine Universität-Düsseldorf, 40225 Düsseldorf, Germany *timo@theochem.uni-duesseldorf.de, kleinermanns@uni-duesseldorf.de

Published in ChemPhysChem (2008)

Abstract

Electronic and vibrational gas phase spectra of 1-methylthymine (1MT) and 1-methyluracil (1MU) and their clusters with water are presented. Mass selective IR/UV double resonance spectra confirm the formation of pyrimidine-water clusters and are compared to calculated vibrational spectra obtained from ab initio calculations. In contrast to Y. He, C. Wu, W. Kong; J. Phys. Chem. A, 2004, 108, 93 we were able to detect 1MT/1MU and their water clusters via resonant two-photon delayed ionization under careful control of the applied water vapor pressure. The long-living dark electronic state of 1MT and 1MU detected by delayed ionization survives hydration and the photostability of 1MT/1MU cannot be attributed solely to hydration. Oxygen coexpansions and crossed beam experiments indicate that the triplet state population is probably small compared to the ^{1}nπ^* and/or hot electronic ground state population. Ab initio theory shows that solvation of 1MT by water does not lead to a substantial modification of the electronic relaxation and quenching of the ^{1}nπ^* state. Relaxation pathways via $^1\pi\pi^*/^1$n$\pi^*$and $^1\pi\pi^*/S_0$ conical intersections and barriers have been identified, but are not significantly altered by hydration.

4.1 Introduction

Absorption of UV radiation by DNA bases can generate mutations which mainly occur at bipyrimidine sites. The major photoproducts are cyclobutane dimers and pyrimidine-(6-4)-pyrimidone adducts. The detailed mechanism of this photoreaction is still unclear – direct reaction in the photoexcited $^1\pi\pi^*$ state [9], a long-living ^{1}nπ^* or triplet state [10, 11] populated by relaxation of the $^1\pi\pi^*$ state and non-concerted product formation in the hot S_0 ground state come into question.

The pyrimidine bases uracil (U) and thymine (T) feature broad absorptions in the gas phase and in solution [12,13,14,15,16,17,18] with an onset at around 275 nm in the vapor spectrum. This first absorption band has been assigned to a $^1\pi\pi^*(S_2 \leftarrow S_0)$ excitation. [14,18,19,20,21] The computed oscillator strength for the $^1n\pi^*(S_1 \leftarrow S_0)$ transition is smaller by a factor of about 10^3, i.e. the state is "dark". [20]

The dominating decay channel of UV-excited thymine is subpicosecond internal conversion from the $^1\pi\pi^*$ state to S_0 [22] creating $\approx 90\,\%$ vibrationally hot T which cools on the ps time scale. [23] Passage to the dark $^1n\pi^*$ state accounts to approximately $10\,\%$ for T in solution. [23] Condensed phase lifetimes of the $^1n\pi^*$ state are typically 30 ps for thymine and $4\times$ longer for thymidine monophosphates (TMP). A small fraction of the $^1n\pi^*$ population is proposed to undergo intersystem crossing to the lowest triplet state T_1 in competition with vibrational cooling. [23]

The triplet states of uracil and thymine have been characterized in considerable detail. Weak phosphorescence of U at 77 K was observed with a maximum at 450 nm (2.76 eV) [12] which compares favorably with the computed vertical $T_1 \rightarrow S_0$ deexcitation energy of ≈ 2.73 eV. [20] The adiabatic transition energy is about 3.2 eV. Phosphorescence is weak because the T_1-S_2 and T_1-S_0 spin-orbit coupling matrix elements are small whereas the significant T_1-S_1 coupling does not lead to enhanced emission because the S_1 state is dark. [20] A triplet lifetime of 75 ms was observed for T at 77 K via phosphorescence [12] in contrast to a unimolecular decay time of 32 μs in deaerated aqueous solutions as measured by nanosecond transient absorption spectroscopy (ns-TAS). [24] Triplet state formation has a low quantum yield < 0.02 in water. [25,26,27,28,26] However, the triplet state is still of interest due to its long lifetime when considering excited state DNA chemistry. TMP has the lowest triplet energy of all nucleotides [29] hence interbase triplet energy transfer would lead to accumulated triplet TMP.

Ns-TAS experiments showed that UV excitation of the single-stranded oligonucleotide $(dT)_{20}$ leads to cyclobutane pyrimidine dimers (CPD) in less than 200 ns, whereas the (6-4) adduct is formed within 4 ms probably via intermediate oxetane. [30] Quantum yields of 0.03 and 0.004 were determined for CPD and (6-4) formation, respectively. Femtosecond time-resolved infrared spectroscopy was used to study the formation of CPD in $(dT)_{18}$ upon 272 nm excitation. [9] The ultrafast appearance of CPD marker bands points to dimer formation in less than 1 ps. It was concluded that T dimerization in DNA is an ultrafast photoreaction and occurs directly in the photoexcited singlet $^1\pi\pi^*$ state. [9]

Spectroscopic experiments in supersonic jets are able to study isolated nucleobases in very detail and may shine more light on the puzzle of thymine's excited states. IR and microwave experiments demonstrated that the diketo tautomer of U and T is the most abundant species in the gas phase, probably the only one. [18,31,32] Resonance Enhanced Two Photon Ionization (R2PI) experiments showed broad and diffuse electronic spectra of U and T which were attributed either to mixing of electronic states or to a large geometry change between the ground and excited electronic states. [18] Kong and coworkers [33] showed that methyl-substituted uracils and thymines can be ionized with tens to hundreds of nanoseconds delay after electronic excitation pointing to the existence of a long-living dark electronically excited state of T in the gas phase which they assigned to a $^1n\pi^*$ state. The lifetime of this dark state depends on the internal energy and the degree of methyl substitution with the general trend that the lifetime increases the more substituted the ring and the longer the pump wavelength is. They observed a decrease of lifetime of the dark state of T and a gradual loss of the ion signal with increasing water content in the jet and concluded that in water solutions the dark state is not populated significantly anymore because fast internal conversion to the electronic ground states dominates. Hence Kong and coworkers inferred that the photostability is not an intrinsic property of the pyrimidine DNA bases but results from their hydration. [33]

The work of Kohler et al. [23] on the corresponding pyrimidine nucleotides and short DNA strands however showed that this state is clearly not quenched in water. We therefore decided to reinvestigate the photophysical behavior of methyl substituted thymine in the gas phase.

In this paper we present the first IR-UV double resonance spectra of hydrated 1-methylthymine (1MT) from which we infer the structures of 1MT-water clusters. In particular we demonstrate that in 1MT (methylation where the sugar sits in the DNA) the dark state survives hydration and delayed ionization is still possible. We observe broad electronic spectra both of 1MT and its water clusters and assign their diffuse appearance to fast internal conversion of the pumped $^1\pi\pi^*$ state. Hence in the isolated and in the hydrated molecule both fast and slow channels are still observable. Extensive high level ab initio calculations are performed to understand this photophysical behavior in detail.

The paper is structured as follows: In the next section we give a detailed account and discussion of our experimental findings and an interpretation of the results based on high-level ab initio quantum-chemical calculations. In the third section we draw conclusions from our investigations and in the final section we describe our experimental and theoretical techniques.

4.2 Results

4.2.1 Spectra

Figure 4.1 shows the time-of-flight mass spectra of 1MT and 1MU and their hydrated clusters at water vapor pressures of 8.2 and 12.3 mbar. Two photon ionization was performed by using 273.2 nm as excitation wavelength and 193 nm for ionization with 40 ns delay between the two laser pulses. At low water vapor pressure (8 mbar) merely the masses of the monomers and the 1:1 water clusters are discernible. We assume that the contribution from fragmenting larger water clusters plays only a minor role at 8 mbar H_2O pressure. The intensity of the monomer ion signal is about 18-times larger than the water cluster ion signal under these conditions. At a higher water pressure of 12.3 mbar, we observed clusters with up to five water molecules and only four times more monomer than 1:1 cluster signal pointing to extensive monomer-water aggregation and considerable fragmentation contributions to the mass spectrum. At still higher water pressures we monitored increasingly larger water clusters and a continuing decrease of monomer signal due to cluster formation down to zero monomer signal. In contrast to Ref. [33], clusters with more than four water molecules were easily detected. When increasing the water vapor pressure, all ion intensities (monomer and cluster masses) decreased and finally vanished, which is probably correlated to a lower ion detection efficiency under these conditions and the formation of "ice-balls" containing 1MT or 1MU. In that case almost all 1MT and 1MU is incorporated into very large clusters. As a consequence, the overall ion-signal intensities on all small cluster masses as well as the monomer mass are reduced. This may have been a problem in earlier measurements [33], as discussed below, in which a water vapor pressure of 23 mbar has been used to generate thymine-water clusters.

The vibronic spectra of jet-cooled dry 1MT and 1MU detected by R2PI are displayed in Figure 9.1a. The spectrum of 1MT reveals a distinct peak at 35880 cm^{-1} (278.7 nm) with an onset at $\sim$35700 cm^{-1} followed by a broad, unstructured band which extends up to 41000 cm^{-1} ($\sim$244 nm). The electronic spectrum of 1MU is essentially unstructured and blue shifted relative to the 1MT spectrum, as expected from the lower degree of methylation, with an onset at $\sim$36400 cm^{-1} (274.7 nm). Figure 9.1b shows the R2PI spectra of 1MT(H_2O) and 1MU(H_2O) clusters detected on the parent masses at low water concentration (8 mbar). As discussed above, we expect preferential formation of small clusters under these experimental conditions. The spectra are again largely unstructured and exhibit only a small shift relative to the monomer spectra. The peak at 35880 cm^{-1} (278.8 nm) in the 1MT spectrum is not discernible anymore in the 1MT/water spectrum.

The IR-UV spectrum of dry 1MT is presented in the top trace of Figure 6.2. The absorption at 3434 cm^{-1} is attributed to the "free" NH stretch vibration. The free NH stretch vibration of 1MU was detected at 3432 cm^{-1} (not shown here). The next trace in Figure 6.2 shows the IR-UV ion dip spectrum recorded at the 1MT$(H_2O)_1$ mass at a still lower water pressure of 6 mbar. Cluster structures containing one and two water molecules calculated at the RIMP2 level and their IR stick spectra are displayed for comparison and assignment of the experimental spectrum. The ground-state geometries obtained from CC2/cc-pVDZ calculations of clusters with up to five water molecules are in close agreement with the RIMP2 structures (compare Figure 4.7).

Closer analysis shows that the experimental spectrum can be ascribed to the NH and OH stretch vibrations of 1MT$(H_2O)_1$ with some contributions from 1MT$(H_2O)_2$ discernible by the shoulder at 3388 cm^{-1}. The IR band of 1MT$(H_2O)_2$ calculated at $\sim$3000 cm^{-1} is readily assigned to the OH stretch vibration in the OH$\cdots$O(H) hydrogen bond and can be expected to be very broad and therefore hard to detect. However, we cannot differentiate between the different isomers of 1MT$(H_2O)_1$ displayed in Figure 6.2 because of the close resemblance of the calculated spectra.

4.2.2 Lifetime Measurements

Figure 4.4a shows the ion signal decay of electronically excited dry 1MT and 1MU as a function of the delay time between the excitation (273.2 nm) and ionization (193 nm) laser pulses. Decay times are 107 ns $\pm$ 30 ns for 1MU and 227 ns $\pm$ 30 ns for 1MT. As already pointed out in Ref. [33] lifetime increases with degree of methylation. This finding is supported by ab initio calculations: For bare 1MT we determine the barrier between the $^1n\pi^*$ minimum and the $^1\pi\pi^*$/ $^1n\pi^*$ conical intersection to be 1.33 eV (details in the following), whereas this barrier is lower than 1 eV for bare uracil. [34]

Figure 4.4b shows the ion signal decay curves in the presence of water, both, on the monomer masses (1MT and 1MU) and on the 1MT/1MU$(H_2O)_1$ cluster masses. The ion signals on the cluster masses show a slow, nearly linear decay. In Figure 4.1, ion signals of clusters with more than four water molecules are readily discernible at a delay time of 40 ns and low water content. At higher water vapor pressures we measured ion signals for clusters with up to eight water molecules. The decay curves of the larger clusters all show the same, nearly linear decay as the 1MT/1MU(H_2O) clusters. Contrary to earlier measurements [33] we do not observe a sharp drop of the ion signal intensities for clusters with more than four water molecules. Instead, the intensities decrease smoothly with increasing cluster size, as in a typical cluster experiment, because more collisions are required to form larger clusters. Figure 4.4 clearly demonstrates that delayed ionization is possible not only for the monomers, but also for their clusters with water. Identical decay curves were obtained when using 213 nm instead of 193 nm for ionization. Kong and coworkers [33] used rather high water vapor pressures in their experiment which leads to a decrease of the ion detection efficiency up to complete disappearance of the delayed ion signal according to our experience.

The ion signal decay curves of the clusters in Figure 4.4b are limited by the time the clusters need to fly out of the ionization volume with the speed of the molecular beam. In our experiment (beam width ≈ 2 mm) it takes about 1.5 μs for all the clusters to fly out of the ionization volume. That time agrees nicely with the extrapolated intersection between the nearly linear decay curves and the time axis in Figure 4.4b. We confirmed this by varying the size of the ionization laser beam. Thus, the long-living dark state of the water clusters decays on a slower time scale

($> 1\,\mu$s) than the size of our observation window.

The lifetime measured on the monomer mass in the presence of water (8 mbar vapor pressure) increases to 218 ns ± 30 ns for 1MU and 363 ns ± 30 ns for 1MT which is attributed to a contribution of water clusters fragmenting to the monomer mass, either during the lifetime of the dark state or after ionization [35, 36].

4.2.3 Oxygen Coexpansion

The identification of the dark state still remains an open question. Possible candidates are the $^1n\pi^*$ state, a low-lying triplet state or the vibrationally hot electronic ground state populated by efficient internal conversion from the excited singlet state. Since ground state triplet oxygen features a low-lying singlet state, efficient 1MT triplet quenching is expected when coexpanding 1MT with oxygen or when crossing a molecular beam of excited 1MT with an oxygen beam. Indeed, quenching of T and TMP triplet-triplet absorption with O_2 was reported to be very effective [24] so that the measurements had to be done in an aqueous solution saturated with N_2 excluding all O_2. Therefore we decided to use oxygen quenching as a check of the triplet character of the dark state.

In a first step, we coexpanded dry 1MT and 1MU in a mixture of 10 % oxygen and 90 % helium. The lifetime did not change within the accuracy of our measurements. In a second approach we established an effusive oxygen beam (100 % O_2), which propagated perpendicular to the molecular beam carrying 1MT and 1MU and again measured the lifetime of the dark state. We compared the ion signal intensities under normal expansion and crossed beam conditions using a delay of 40 ns between the excitation and ionization laser pulses. Crossed beam experiments with nitrogen and argon were performed as a reference for the attenuation of the molecular beam due to elastic and inelastic collisions in the crossed beam setup. Comparing the ion intensities gives valuable information, even if the quenching process occurs on a time scale below the detection limit of our apparatus (10 ns). In that case quenching is still observable by a decreased ion signal intensity.

Indeed, we observed a small intensity decrease. However, the effect was similarly large in collisions with nitrogen or argon and can therefore not be attributed to specific triplet quenching. Therefore the triplet state population is probably small compared to the population of the $^1n\pi^*$ or hot vibrational ground state. Kong et al. observed weak fluorescence of 1,3-dimethyluracil with similar lifetimes as the decay times of the delayed ionization signal so that the hot S_0 state seems to us a less likely candidate for the dark state than the $^1n\pi^*$ state. Note, however, that delayed fluorescence in which a low-lying excited singlet state becomes populated by a thermally activated radiationless transition from the first excited triplet state is a well-known phenomenon in condensed-phase spectroscopy. [9, 10]

The variance in decay times for bare and hydrated 1MT might be related to the relative energies of the involved electronic states. To shine more light on the change of the lowest $^1n\pi^*$ and $^1\pi\pi^*$ singlet and triplet states with hydration we performed extensive high-level ab initio calculations. We specifically focus on the following questions: Is a direct decay from the bright $^1\pi\pi^*$ state to the electronic ground state enhanced by hydration? Does hydration significantly affect the energy barrier for an alternative decay via the dark state?

Our single-point/adiabatic CC2 and LIIC CASSCF/CASPT2 calculations (Figures 4.10, 4.11 and 4.12, supplementary material) enable us to construct an energy level diagram for bare 1MT as shown in Figure 4.5. In accord with previous investigations [37, 38, 39] we find conical intersections $^1\pi\pi^*/^1n\pi^*$ (CI_1) and $^1\pi\pi^*/S_0$ (CI_2). Furthermore, the two excited states in question exhibit minima on the potential hypersurfaces. The structures at the conical intersections are shown in Figure 4.6. The structures of the 1MT-water clusters in the electronic ground and $^1n\pi^*$ state optimized at the CC2/cc-pVDZ level are displayed in Figure 4.7 and Figure 4.9 (supplementary material).

Our calculations showed that LIIC paths for bare 1MT and singly-hydrated 1MT are essentially the same. We may restrict the discussion to the variance of relative energies due to hydration at crucial points on the hypersurfaces, i.e., at the respective minima and conical intersections. From our most accurate calculations (CC2/aug-cc-pVTZ) we determined a $^1n\pi^*$ adiabatic excitation energy of 3.73 eV and a barrier between the minimum of $^1n\pi^*$ and CI_1 of 1.33 eV for bare 1MT. Upon microhydration with one water molecule, this barrier is lowered to 1.09 eV and the adiabatic energy is increased to 3.88 eV. Due to these findings we expect a higher probability of $^1\pi\pi^*/^1n\pi^*$ internal conversion upon hydration with one water molecule than for bare 1MT.

To address the question of whether increased microhydration might further lower this barrier or lead to a dramatically enhanced direct relaxation from $^1\pi\pi^*$ to S_0 through CI_2 we have carried out a series of CC2/aug-cc-pVTZ single-point calculations. The results are shown in Figure 4.8 and Figure 4.13 (supplementary material) for two points, the equilibrium geometry of the ground state and the equilibrium geometry of the $^1n\pi^*$ excited state. Interestingly, we observe the trend that the $^1n\pi^*$ state is destabilized energetically relative to the $^1\pi\pi^*$ state for structures with one and two water molecules. This points to both an enhanced direct relaxation and a reduced population of the $^1n\pi^*$ state. The trend, however, is broken at the time when the third and subsequent water molecules are attached to the system. In the cluster with five water molecules only a minute deviation of the relative energies from those of bare 1MT remains. We ascribe this energetic reconstitution to a recovery of partial symmetry at the substitution sites (where the hydrogen bonds perturb the electronic valence distribution) as soon as the third water molecule enters the cluster. The preferred formation of water clusters around the oxygen binding sites is shown in Figure 4.7 where also the symmetry argument becomes obvious. Thus, a substantial modification of the relaxation pathways of 1MT through microhydration is not observed and a claim that solvation by water would lead to a quenching of the dark state cannot be supported by ab initio theory.

4.3 Discussion

In this work, we tried to shine more light on the decay mechanisms of 1MT/1MU and their correlation to aggregation with water. We are, to our knowledge, the first to publish R2PI spectra of 1MT/1MU and their water clusters. IR/UV double resonance experiments confirmed the formation of 1MT/1MU(H_2O) clusters, in which the water molecule forms a bridge between the NH bond and a neighboring C=O group. Ab initio calculations nicely reproduce the experimental spectra. The infrared spectra also show, that fragmentation of larger water clusters is minimized at low water concentrations in the supersonic expansion. In contrast to the work published by Kong and coworkers [33], we were able to detect delayed ionization of 1MT/1MU-water clusters up $1\,\mu s$ after excitation under careful control of the applied water vapor pressure. Even aggregates carrying more than five water molecules were detected. In addition, our ab initio calculations show that the energetics of the excited states and the de-excitation pathways are not significantly altered by hydration.

The decay curves of the bare monomers (1MT and 1MU) in Figure 4.4a show an exponential decay much faster than our upper detection limit of about $1.5\,\mu s$. We can only detect molecules (or clusters) that are ionizable at the wavelength of our ionization laser (193 or 213 nm). This means that on the time scale of our experiment 1MT and 1MU relax from a higher to a lower state, which is not ionizable anymore. The pyrimidine bases thymine and uracil and their derivatives undergo ultra fast (sub-picosecond time scale) internal conversion from the initially excited

$^1\pi\pi^*$ state to the vibrationally excited ("hot") S_0 ground state and to some extent to the optically dark $^1n\pi^*$ state. [22, 23] For the moment, let us assume that the ionization laser also probes the hot electronic ground state. IVR would spread the energy over the vibrational degrees of freedom on a time scale much faster than our lower detection limit of 10 ns. However, since the gas phase experiment takes place in the collision free region of a skimmed molecular beam, the energy will stay in the molecule as long as the radiation lifetime of the vibrational modes ($> 1\,\mu s$). In that case the molecule could be ionized at any time with the same probability, resulting in a linear decay as the molecules fly out of the ionization volume.

This is clearly not the case for the bare monomers in Figure 4.4a. After 600 ns only a small fraction of the monomers can be ionized, indicating that the hot ground state either cannot be ionized at 193 nm or that the ionization probability is too low for detection due to an unfavorable Franck-Condon overlap. We therefore attribute the exponential decay of the monomer ion signal to the true relaxation from an electronically excited state (the dark state) to the S_0 ground state.

By contrast, the 1MT- and 1MU-water clusters show a linear decay of the ion signal, meaning that the probed state has a longer lifetime than our upper detection limit. However, we have no reason to believe that the hot ground state of the clusters has a higher ionization probability than the monomer nor do the energetics change significantly between the monomer and clusters, which could have explained a faster decay to the hot ground state. To the contrary, the higher number of vibrational degrees of freedom is likely to facilitate the energy dissipation via IVR. In addition, vibrationally hot clusters tend to fragment by which they will loose energy (evaporative cooling). Therefore, we attribute the slow decay of the cluster ion signal also to the relaxation of an electronically excited state, which is most likely the optically dark $^1n\pi^*$ state. Our upper detection limit is then only a lower limit for the dark state lifetime ($> 1\,\mu s$).

Kohler and coworkers [40] showed that hydrogen bonds between nucleobases and the solvent strongly enhance vibrational cooling, allowing the nucleobases to dissipate vibrational excess energy to the solvent more effectively. The additional hydrogen bonds between the ribosyl group and the water solvent are responsible for the four-fold larger lifetime of the TMP and UMP nucleotides compared to the nucleobases T and U in solution. [23] Similar effects might be responsible for the increased dark state lifetime of the 1MT- and 1MU-water clusters compared to the isolated nucleobases in the gas phase. A more effective energy dissipation of the hydrated monomers as well as cluster fragmentation during the lifetime of the dark state (evaporative cooling) might lead to a slower passage of the $^1n\pi^*/S_0$ barrier and an increase of the dark state lifetime.

4.4 Summary

Our experimental results demonstrate that hydration does not accelerate internal conversion to the electronic ground state to a degree that the dark state is quenched as pointed out in Ref [33]. Accordingly, the thesis of Kong et al. [33] that the photostability is not an intrinsic property of the pyrimidine bases but results from their hydration is not confirmed for 1MT and 1MU. To the contrary, we propose an increase of the dark state lifetime of 1MT- and 1MU-water clusters in the gas phase. Our results agree with the observation of a long-living dark state upon UV excitation of T and U nucleotides in aqueous solutions. [23] Ab initio theory shows that solvation of 1MT by water does not lead to a substantial modification of excited-state energetics and therefore electronic relaxation and quenching of the $^1n\pi^*$ state. Coexpansions and crossed beam experiments with oxygen showed no measurable decrease of the dark state lifetime, so that the triplet state population is probably small compared to the $^1n\pi^*$ population. Further experimental and theoretical work is directed towards identification of the dark state via vibrational spectroscopy and towards clarification of a possible connection of the dark state to the dimerization of thymine.

4.5 Experimental and Theoretical Section

The experimental setup has been described in detail elsewhere. [35] A mixture of helium and thymine derivatives at 150 to 170 °C is expanded through the 300 μm orifice of a pulsed nozzle (General Valve). The skimmed molecular beam (Beam Dynamics Skimmer, 1 mm orifice) crosses the collinear laser beams at right angle. The ions are extracted in a modified Wiley-McLaren type time-of-flight (TOF) spectrometer perpendicular to the molecular and laser beams.

1-Methylthymine (1MT) and 1-methyluracil (1MU) were purchased from Sigma-Aldrich and used without further purification. Water clusters were formed by bubbling helium through water prior to the expansion. The water content was controlled by the temperature dependence of the water vapor pressure. Different vapor pressures of 6 mbar (273 K), 8.2 mbar (277 K) and 12.3 mbar (283 K) were investigated.

R2PI (resonance enhanced two-photon ionization) spectra were obtained by exciting the molecules to the first optically accessible electronic state by the frequency doubled output of a Nd:YAG (Spectra Physics, GCR 170) pumped dye laser (LAS, LDL 205) operated with Fluoresceine 27, Coumarine 307 and Coumarine 153. The molecules were ionized by delayed ionization using the 193 nm output of an ArF excimer laser (Neweks, PSX-501-2) or the fifth harmonic of a Nd:YAG laser (213 nm, Innolas, Spitlight 600). Longer ionization wavelengths reduce the excess energy in the ions and thus fragmentation of the 1MT clusters. For the IR/UV double resonance spectra an infrared (IR) laser pulse was fired 36 ns prior to the UV excitation laser. IR light (3000–4000cm^{-1}) was generated by an IR-OPO/OPA setup (LaserVision), pumped by the fundamental of a Nd:YAG laser (Spectra Physics, Indi 40-10). IR laser frequencies were calibrated by recording an ammonia vapor spectrum.

All calculations were performed with the quantum chemistry program packages TURBOMOLE [41] and MOLPRO [42]. We carried out Resolution-of-the-Identity (RI)-Coupled-Cluster (CC) 2 [43, 44] geometry optimization [45] of ground- and excited states and RI-CC2 in combination with linear response theory for vertical single-point calculations. We used Dunning's correlation-consistent basis sets cc-pVDZ (C, N, O: 9s4p1d/3s2p1d; H: 4s1p/2s1p) for geometry optimizations and aug-cc-pVTZ (C, N, O: 11s6p3d2f/5s4p3d2f; H: 6s3p2d/4s3p2d) for excitation energies which have provided results of high accuracy in previous CC2 calculations on nucleic acid bases [46]. Geometries were optimized without symmetry constraints. The optimization of conical intersections and the calculation of Linear Interpolation in Internal Coordinates (LIIC) paths have been carried out with state-averaged CASSCF/CASPT2 (Complete-Active-Space Self-Consistent-Field/Complete-Active-Space Perturbation Theory to 2nd order) methods [47,48,49] and Pople's 6-31G* (C, N, O: 10s4p1d/3s2p1d; H: 4s/2s) basis set [50]. The optimization of conical intersections was carried out with the CASSCF method with an active space of 10 electrons in 8 molecular orbitals (3 π, 2 n, 2 π^* and 1 σ^* orbitals) and the 6-31G* basis set. We obtained starting geometries for conical intersection optimization by slightly deforming the ring in order to destabilize the electronic ground state. In all calculations we retained a frozen core comprised by all heavy-atom 1s electrons.

Acknowledgement:

The authors thank the Deutsche Forschungsgemeinschaft (SFB 663) for financial support.

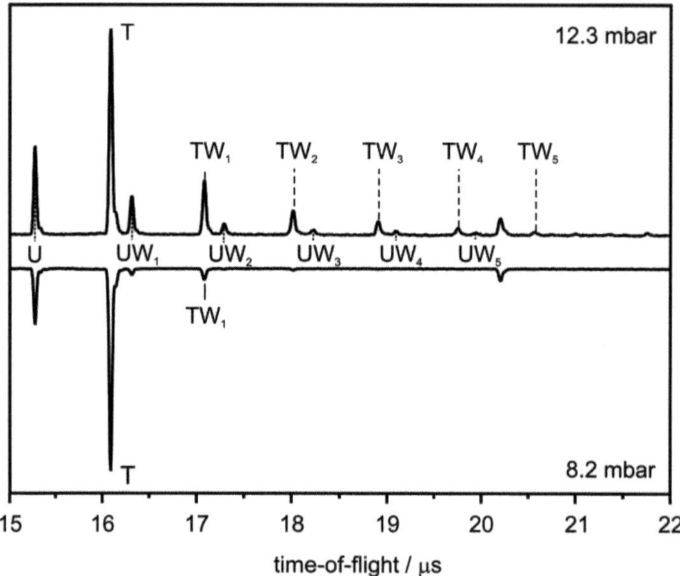

Figure 4.1: *Mass spectra of 1MT and 1MU (abbreviated here as T and U) and their water clusters at water vapor pressures of 12.3 mbar (top trace) and 8.2 mbar (lower trace). The ion signal of the lower spectrum has been inverted for easier comparison. Excitation at 273.2 nm, ionization at 193 nm; 40 ns delay between the excitation and ionization laser pulses.*

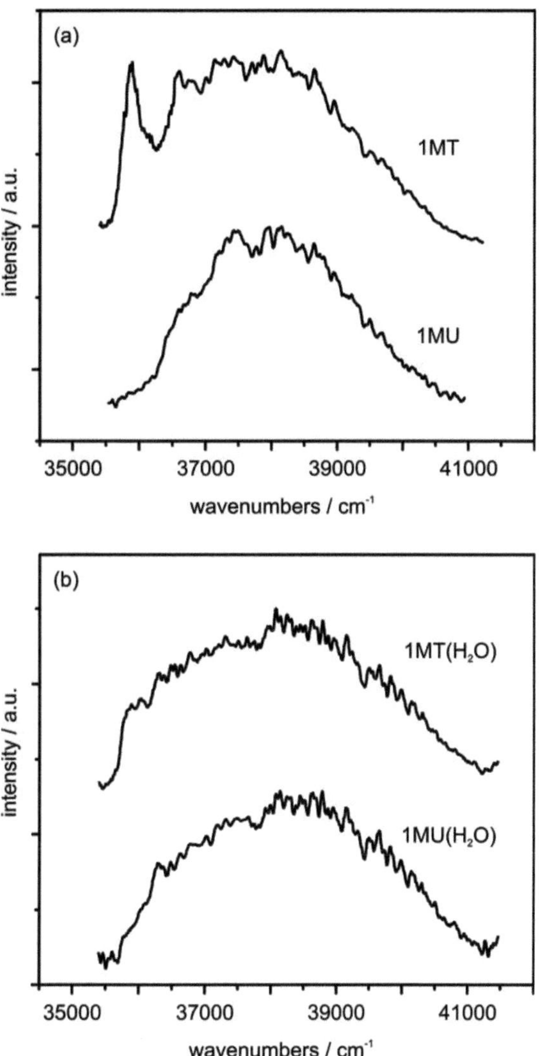

Figure 4.2: *a) Vibronic spectra of 1MT and 1MU monomers (without addition of water) detected by R2PI. b) R2PI spectra of 1MT/1MU-water clusters detected at the 1MT/1MU(H_2O)$_1$ mass at low water concentration (8 mbar). Ionization at 193 nm; 40 ns delay between the excitation (300 μJ for clusters and 120 μJ for monomers) and ionization (30 μJ) laser pulses.*

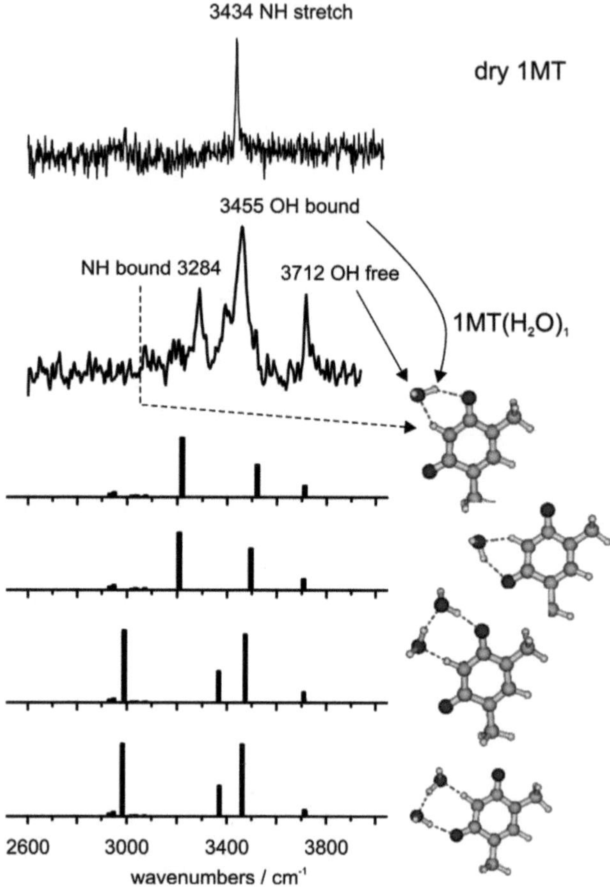

Figure 4.3: *Top trace: IR-UV spectrum of dry 1MT. Lower trace: IR-UV spectrum of 1MT-water detected at the 1MT(H_2O)$_1$ mass at low water concentration (6 mbar vapor pressure). Excitation at 273.3 nm, ionization at 213 nm. Displayed are the most stable structures of 1MT(H_2O)$_{1,2}$ and their infrared spectra calculated at the RIMP2/ccpVDZ level. The vibrational frequencies were scaled by a factor of 0.9502 to match the calculated free NH-Stretch frequency of the 1MT monomer to its experimental value.*

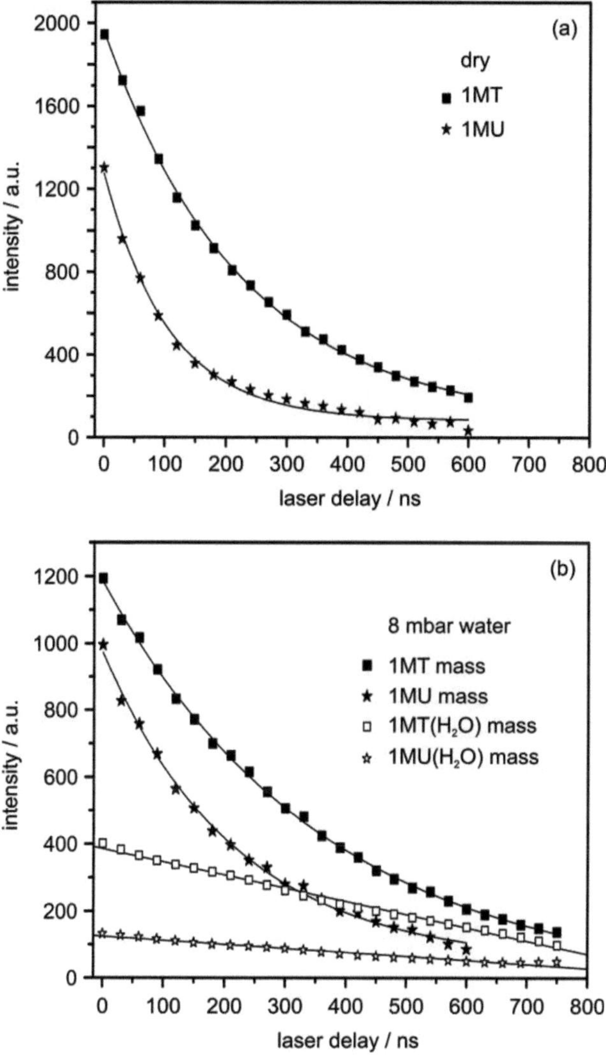

Figure 4.4: *Ion signal decay of dry 1MT and 1MU (a) and of hydrated 1MT and 1MU (b). Excitation at 273.2 nm (120 µJ) and ionization at 193 nm (30 µJ) with a variable time delay between the excitation and ionization laser pulses. Water vapor pressure: 8 mbar (4° C).*

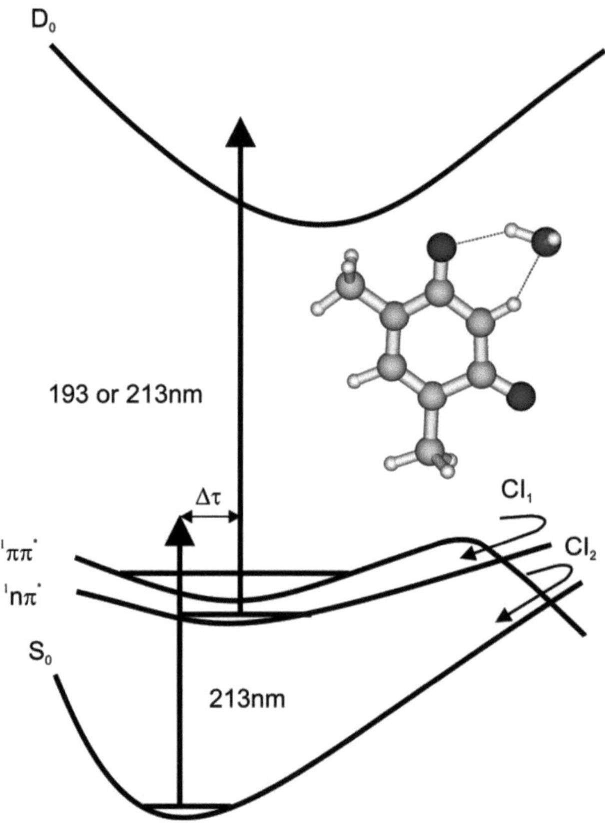

Figure 4.5: *Schematic representation of possible relaxation pathways in* $1MT(H_2O)_1$ *(triplet states excluded).*

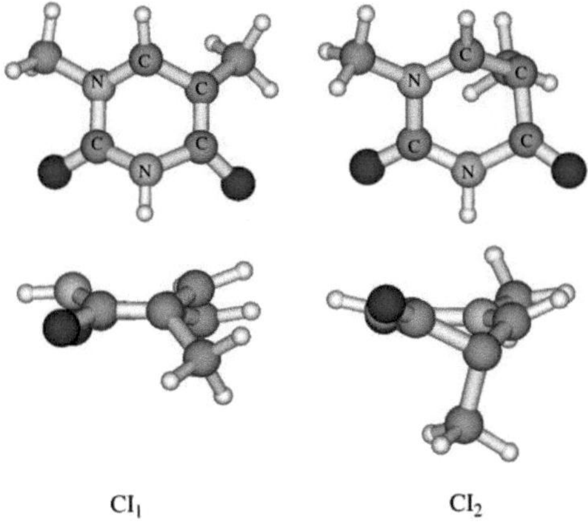

Figure 4.6: *Geometries of the conical intersections in 1MT optimized at the CASSCF(10,8)/6-31G* level: CI_1 is the intersection between $^1\pi\pi^*$ and $^1n\pi^*$ adiabatic states and CI_2 is the intersection between $^1\pi\pi^*$ and ground adiabatic states.*

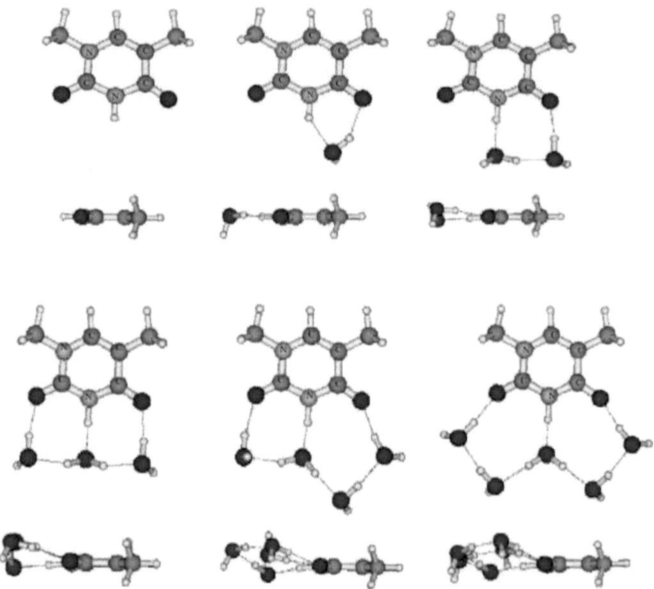

Figure 4.7: *Geometries of ground state 1MT and 1MT-water clusters optimized at the CC2/cc-pVDZ level.*

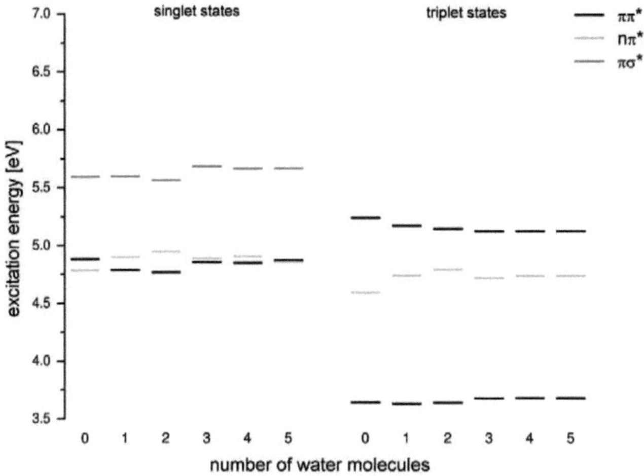

Figure 4.8: *Vertical electronic excitation energies calculated at the ground state geometry at CC2/aug-cc-pVTZ level (the three lowest states). All structures have been optimized at the CC2/cc-pVDZ level.*

Supplementary Material

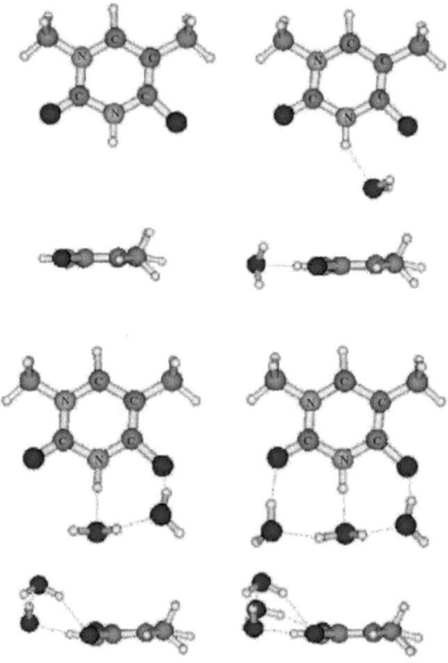

Figure 4.9: *Geometry of the $^1n\pi^*$ state of 1MT and 1MT-water clusters optimized at the CC2/cc-pVDZ level.*

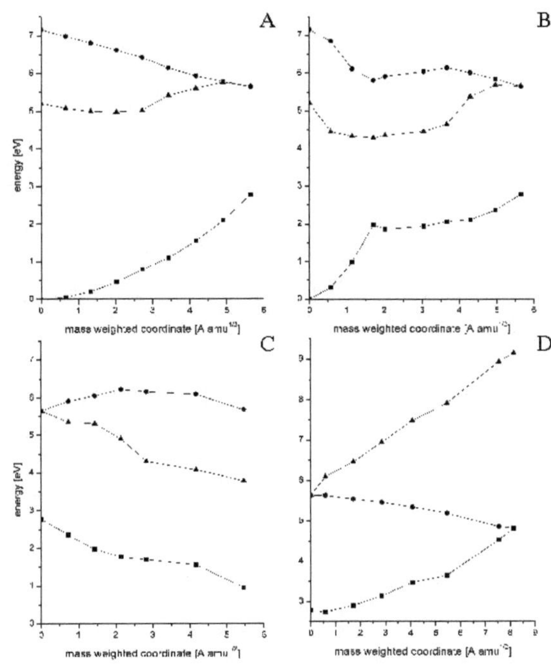

Figure 4.10: *Potentional energy profiles of the ground (squares), $^1n\pi^*$ (triangles) and $^1\pi\pi^*$ state (circles) of 1-methylthymine, calculated at the CASSCF(10,8)/6-31G* level of theory along the LIIC reaction path: A from the equilibrium geometry of the ground state to the CI_1; B from the equilibrium geometry of the ground state to the minimum of the $^1\pi\pi^*$ state and to the CI_1; C from the CI_1 to the minimum of the $^1n\pi^*$ state; D from the CI_1 to the CI_2*

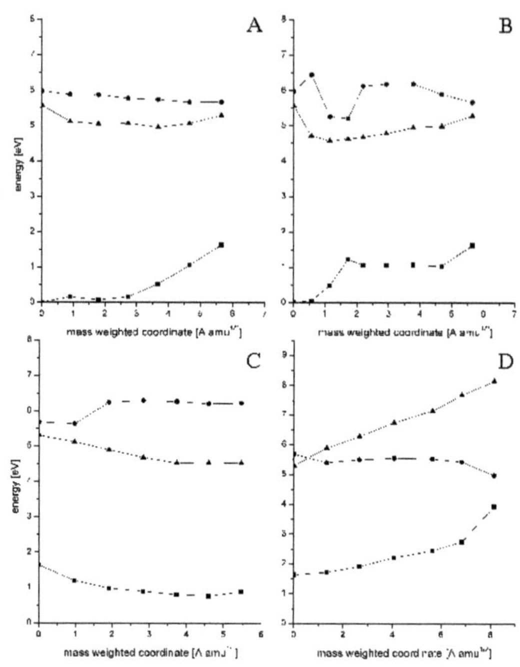

Figure 4.11: *Potentional energy profiles of the ground (squares), $^1n\pi^*$ (triangles) and $^1\pi\pi^*$ state (cirles) of 1-methylthymine, calculated at the CASPT2(8,7)/6-31G* level of theory along the LIIC reaction path: A from the equilibrium geometry of the ground state to the CI_1; B from the equilibrium geometry of the ground state to the minimum of the $^1\pi\pi^*$ state and to the CI_1; C from the CI_1 to the minimum of the $^1n\pi^*$ state; D from the CI_1 to the CI_2*

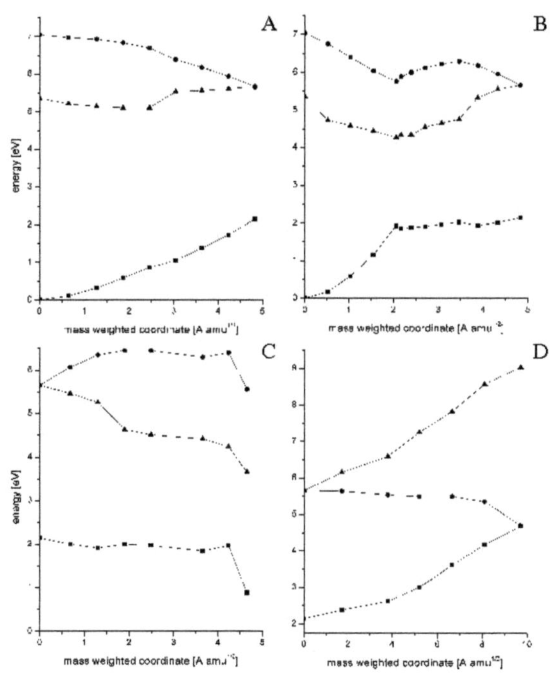

Figure 4.12: *Potentional energy profiles of the ground (squares), $^1n\pi^*$ (triangles) and $^1\pi\pi^*$ state (cirles) of 1-methylthymine · H_2O, calculated at the CASSCF(10,8)/6-31G* level of theory along the LIIC reaction path: A from the equilibrium geometry of the ground state to the CI_1; B from the equilibrium geometry of the ground state to the minimum of the $^1\pi\pi^*$ state and to the CI_1; C from the CI_1 to the minimum of the $^1n\pi^*$ state; D from the CI_1 to the CI_2*

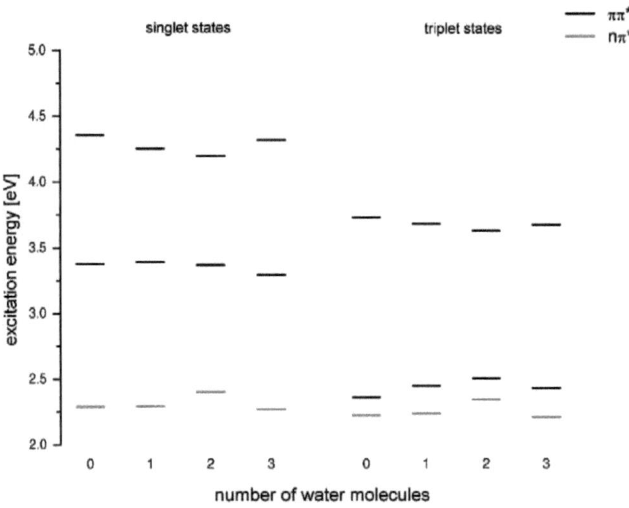

Figure 4.13: *Vertical electronic excitation energies calculated at the geometry of the $^1n\pi^*$ state at CC2/aug-cc-pVTZ level (the three lowest states). All structures have been optimized at the CC2/cc-pVDZ level.*

5 The Dark State of 1-Methylthymine

The nature of the dark intermediate state, which was already discussed in section 4, is of great importance for unveiling the pathways that lead to cyclobutane pyrimidine dimer (CPD) formation.

In order to shine more light on this spectroscopically dark state, the molecule is pumped into it's electronic S_1 state, from which it relaxes to the dark state on an ultrafast timescale. The molecule is then probed by a photon which energy is well below the ionization potential. If the dark state is a triplet state, the molecules could be excited to higher triplet states and by this triplet triplet absorptions could be probed by delayed 1+1' REMPI spectroscopy.

The pump UV-laser was fixed at 273 nm and the probe laser was scanned over a range of 29600 to 40800 cm^{-1}. The used laser dyes were DCM, Rhodamine 101 + Rhodamine B, Pyrromethane 597, Pyrromethane 580, Coumarine 153 and Coumarine 307. The two color REMPI spectrum is presented in figure 5.1.

Unfortunately, the spectrum just shows a broad increase in ion signal at the blue end of the spectral range, which emerges from the weak one color signal produced by the scanned laser. There was no two color signal, which would indicate efficient triplet-triplet (in the range of 350 to 300 nm) absorption.

Calculations simulating the vertical excitation spectrum from the $^3\pi\pi^*$ state and geometry only show very low oscillator strengths for triplet triplet absorptions, which has to be the bottle-neck of this experiment.

The vertical excitation frequencies calculated by Mihajlo Etinski are presented in table 5.1

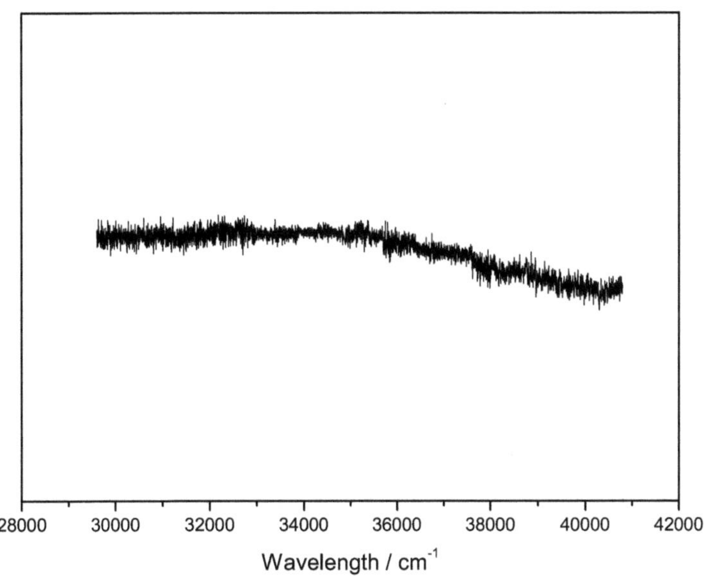

Figure 5.1: *Two color REMPI scan of the 1MT triplet and singlet transition range. Laser delay = 15ns, laser energies are 500 μJ at 273 nm and 1-2 mJ at the probe wavelength*

Transition	Oscillator strength	Wavelength [nm]
$2a \rightarrow 7a$	0.00003	418
$2a \rightarrow 8a$	0.00424	361
$2a \rightarrow 9a$	0.00363	330
$2a \rightarrow 10a$	0.00012	325
$2a \rightarrow 11a$	0.00004	321
$2a \rightarrow 12a$	0.00016	312
$2a \rightarrow 13a$	0.00036	296
$2a \rightarrow 14a$	0.00567	291
$2a \rightarrow 15a$	0.00196	289
$2a \rightarrow 16a$	0.00243	287
$2a \rightarrow 17a$	0.00006	280
$2a \rightarrow 18a$	0.00185	276
$2a \rightarrow 19a$	0.00211	269
$2a \rightarrow 20a$	0.00113	266

Table 5.1: *Vertical excitation frequencies starting from the T_1 ($^3\pi\pi^*$) geometry*

6 The Pyrazine Dimers

IR-UV Double Resonance Spectra of Pyrazine Dimers: Competition between CH$\cdots\pi$, $\pi\cdots\pi$ and CH$\cdots$N interactions

M. Busker, Yuriy N. Svartsov, Th. Häber, K. Kleinermanns

[1] Institut für Physikalische Chemie und Elektrochemie 1, Heinrich-Heine Universität-Düsseldorf, 40225 Düsseldorf, Germany

*kleinermanns@uni-duesseldorf.de

Published in CPL (2008)

6.1 Abstract

We present size- and isomer-selective IR-UV double resonance spectra of two pyrazine dimer isomers. The most stable isomer has a planar structure, stabilized by two CH$\cdots$N contacts. The other isomer has a stacked, cross-displaced structure. Our assignment is supported by B3LYP-D calculations. RI-MP2 calculations tend to overestimate the stability of the stacked and T-shaped isomers.

6.1.1 Introduction

Weakly polar CH groups are non-conventional hydrogen bond donors whereas aromatic π-systems and triple bonds can act as weak hydrogen bond acceptors. Despite their weakness CH$\cdots$O, N, π interactions are supposed to significantly contribute to the stabilization of supramolecular aggregates, crystal packing, molecular recognition and folding of proteins. CH$\cdots$O bonds have been thoroughly investigated in the past. [51, 52] They involve an electrostatic interaction, operate over distances beyond the van der Waals limit and are directional. However, much less is known about CH$\cdots$N interactions (see Ref. [53] and reference therein).

Crystallization of aromatic azacycles like pyridine (Pd) and pyrazine (Pz) and their derivatives enabled the analysis of CH$\cdots$N networks and supramolecular CH$\cdots$N synthons in stoichiometrically defined complexes by X-ray methods. [53, 54, 55] On the other hand, supersonic expansions allow for the formation of small molecular complexes (clusters) under size-controlled conditions which can be compared to the smallest building blocks of the crystal network. The infrared (IR) spectra of these gas phase complexes are especially sensitive to CH$\cdots$X interactions because of the induced spectral shifts of the CH stretching vibrations. In combination with resonance enhanced 2-photon ionization (R2PI) and time-of-flight mass detection this provides a powerful tool for mass- and isomer-selective detection of the infrared spectra of different clusters.

While the benzene dimer has been studied by several groups, both theoretically and experimentally, [56, 57, 58, 59, 60, 61] very little is known about the dimers of the analogous azacycles pyridine, pyrazine and pyrimidine. Two-color R2PI spectra of pyrazine and pyrimidine at various ionization energies revealed several different isomers in supersonic jets. [62, 63] In the case of pyrazine two isomers have been identified and assigned to a T-shaped and a planar, doubly CH$\cdots$N bridged dimer based on calculations employing a Lennard-Jones as well as a hydrogen bonding potential. Recently Mishra and Sathyamurthy published MP2, MP4 and CCSD(T) calculations of several pyrazine dimer isomers. [64] They found the cross displaced stacked dimer (C_S symmetry) to be the most stable at the MP2 level, while at the MP4 and CCSD(T) levels the T-shaped dimer with the N-atom of the stem pointing to the top becomes the most stable structure.

In this paper we report size and isomer selected infrared spectra of pyrazine dimers (Pz_2) and compare them to vibrational spectra of various possible dimer structures calculated at the RI-MP2 and DFT-D levels of theory, utilizing the TZVP and TZVPP basis sets.

6.2 Experiment

The basic principles of our IR-UV experimental setup were described in detail elsewhere. [65, 66, 67] A gas mixture of about 0.2 % pyrazine (Acros Organics, > 99 %) in helium (Air Liquide, 5.0) was expanded through the 300 μm orifice of a pulsed valve (Series 9, General Valve) at a stagnation pressure of 2 bar. The molecules are cooled down to a few Kelvin in the adiabatic expansion where they form clusters.

The skimmed molecular beam (skimmer diameter 2 mm) crosses the UV excitation laser (LAS, frequency doubled, 10 μJ/pulse) and the ionization laser (FL 2002, Lambda Physics, frequency doubled, 120 μJ/pulse) at right angle inside the ion extraction region of a linear time-of-flight (TOF) mass spectrometer in Wiley-McLaren configuration. The UV excitation and ionization lasers are spatially and temporally overlapped. Resonant two photon ionization (R2PI) spectra are recorded by scanning the frequency of the excitation laser between 30820 and 30940 cm^{-1}, while the ionization laser is kept at a fixed wavelength of 233 nm ($\approx$ 42918 cm^{-1}) or 235.5 nm ($\approx$ 42463 cm^{-1}). Since the pyrazine monomer requires an ionization laser energy of more than 44000 cm^{-1}, the strong monomer signal is effectively suppressed by the selected ionization wavelengths. [62,63] In addition, the two-color scheme minimizes excess energy in the ions and reduces the amount of fragmentation after ionization.

For the IR/UV double resonance experiments a pulsed IR laser beam (burn laser, 0.2 cm^{-1} resolution) is aligned collinear to the UV excitation beam (probe laser) and fired 150 ns before the latter. The IR laser frequency is scanned over the vibrational transitions and removes vibrational ground state population if resonant, while the UV excitation laser is kept at a frequency resonant with a vibronic transition of a single cluster isomer. By monitoring the ion mass signal as a function of IR frequency, cluster mass and isomer selective infrared spectra, detected as ion dips, can be obtained. IR laser light is generated by a two-stage setup [68], producing infrared radiation between 2800 and 4000 cm^{-1}(10 Hz, 4 mJ/pulse). The rovibrational transitions of the NH stretching vibrations of ammonia were used for frequency calibration.

The RI-MP2 [69,70] and DFT-D [71,72] calculations were performed with the TURBOMOLE V5.10 program package. [41] Calculated harmonic vibrational frequencies were scaled by 0.9694 (B3LYP-D/TZVP) to match the experimental absorption maximum of the pyrazine monomer C–H stretching vibration at 3070 cm^{-1} [73]. Dimerization energies are not BSSE corrected.

6.3 Results and Discussion

Figure 9.1 shows the two-color R2PI spectra of the pyrazine dimer $(Pz)_2$ in the region of the pyrazine S_1 origin at 30876 cm^{-1}. [74] The spectra were obtained on the pyrazine dimer mass channel and at different ionization energies. At lower ionization energies (bottom trace) we observed bands at −26, −11, −6, +34 and +51 cm^{-1} relative to the monomer origin. With increasing ionization laser frequency (top trace) the bands at −11, +12 and +26 cm^{-1} gain intensity compared to the others (arrows in Figure 9.1), indicating the presence of at least two different isomers. We did not observe any considerable fragmentation of the cluster ions. Below an ionization frequency of 42000 cm^{-1} no dimer ion signals were observed. The ionization energies used to obtain the spectra in Figure 9.1 are still below the ionization threshold of the monomer which is above 44000 cm^{-1}. Overall the R2PI spectra are in agreement with earlier measurements. [62] From now on we will call the isomer with the lower ionization threshold "isomer 1" and that with the higher ionization threshold "isomer 2".

It should be noted that the dimer spectra could be obtained only when the UV excitation and ionization laser pulses (≈10 ns pulse width) were perfectly overlapped in time. No ion signals were observed when delaying the ionization laser by as little as 10 ns with respect to the excitation laser pulse. Thus the excited state lifetime seems to be much shorter than for the monomer, either due to more efficient non-radiative decay channels or due to fragmentation within the electronically excited state.

Figure 6.2 shows the IR-UV double resonance spectra of isomers 1 and 2 (top two traces) between 3000 and 3125 cm^{-1} with the UV excitation laser tuned to the respective 0^0_0 origins at 30848 cm^{-1} (−27) and 30863 cm^{-1} (−12). Both isomers were

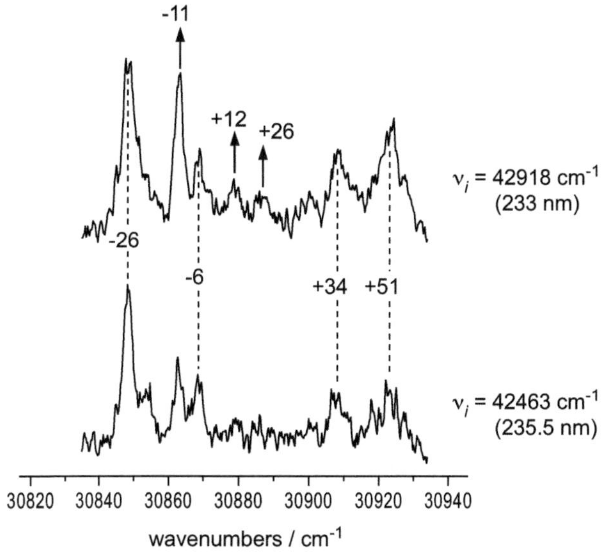

Figure 6.1: *Two color R2PI spectra of pyrazine dimers (Pz_2) in the region of the 0^0_0 band of pyrazine. The spectra were measured at different ionization laser energies of $\nu_i = 42463\,cm^{-1}$ (bottom) and $42918\,cm^{-1}$ (top). At higher ionization energies several bands gain intensity (arrows), indicating the presence of a second isomer. [63, 62] Frequency shifts are given relative to the monomer 0^0_0 band at $30876\,cm^{-1}$. [74]*

ionized at 233 nm. Also shown is the IR-UV spectrum of the pyrazine monomer (bottom) obtained at its S_1 origin ($30876\,cm^{-1}$ [74]) and an ionization wavelength of 193 nm. The monomer shows a single absorption at $3070\,cm^{-1}$ in agreement with the literature. [73, 75]

Interestingly, the infrared spectra of the two dimers are completely different. Isomer 2 absorbs almost at the same frequency as the monomer, with a small shoulder on its low frequency side. The similarity to the monomer spectrum implies that the CH bonds of the absorbing UV chromophore are most likely not directly affected by dimerization. That would be the case in a sandwiched or parallel-displaced geometry. Alternatively, it could be a T-shaped structure with only a weak CH$\cdots\pi$ interaction like the benzene dimer. [57, 58, 59, 60] In that case the CH stretching vibrations are only shifted by 2-3 cm^{-1} compared to the benzene monomer.

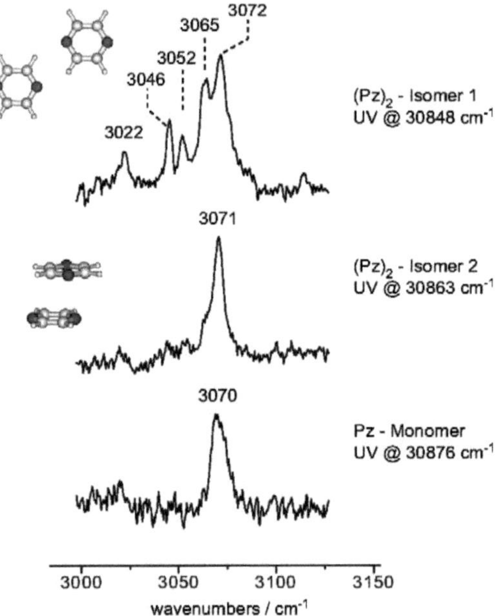

Figure 6.2: *IR-UV double resonance spectra of the two pyrazine dimer isomers (Pz_2, top two traces) and the pyrazine monomer (bottom). The molecules/clusters were excited at their respective 0^0_0 origins and ionized at 233 nm (dimers) or 193 nm (monomer).*

By contrast the infrared spectrum of isomer 1 (top trace in Figure 6.2) shows five reproducible absorptions at 3022, 3046, 3052, 3065 and 3072 cm^{-1}. As mentioned earlier, no fragmentation of the dimer or of larger clusters has been detected. Therefore all observed bands belong to isomer 1. The number of bands and the observed frequency shifts, relative to the monomer absorption point to a direct involvement of the CH bonds of one or of both pyrazine molecules in $CH \cdots \pi$ or, more likely, $CH \cdots N$ interactions. We tend to rule out $CH \cdots \pi$ interactions because the effect on the CH stretching frequencies is rather small, as was demonstrated for the benzene dimer. [76]

Our conclusions are confirmed by B3LYP-D and RI-MP2 calculations of several isomeric pyrazine dimers. Figure 6.3 graphically illustrates the optimized structures of all stable isomers we have found, sorted with increasing B3LYP-D energy. The dimerization energies $E(dimer) - 2E(monomer)$ (including zero-point energy, ZPE) for both methods and for the TZVP and TZVPP basis sets are listed in Table 6.1. Dimerization energies provide a direct comparison between different methods and basis sets. The MP2 results in Table 6.1 and Figure 6.4 are in agreement with the work of Mishra and Sathyamurthy [64], but they fixed the molecular geometry of the monomer units in the dimer calculations, which is not adequate for such weakly bound systems. [77]

The crossed-displaced, stacked dimer (isomer B in Figure 6.3) is part of the so called S22 training set [78] and, consequently, has been the subject of several benchmark studies, including DFT-D calculations. [79, 80] They all predict DFT-D stabilization energies in good agreement with the CCSD(T) benchmark value, while MP2

Table 6.1: *Dimerization energies [E(dimer) − 2E(monomer)] (including ZPE, in kJ mol^{-1}) of various pyrazine dimer structures at different levels of theory. See Fig. 6.3 for a graphical representation of the structures.*

	B3LYP-D		RI-MP2	
Isomer	TZVP	TZVPP	TZVP	TZVPP
A	-17.8	-16.7	-14.5	-14.2
B	-14.8	-14.6	-30.5	-24.0
C	-13.4	-12.9	–[a]	-15.8
D	-12.8	–[a]	-20.5	–[a]
E	-10.3	(-10.7)[b]	-12.7	-10.4
F	-9.7	(-10.5)[b]	-14.3	(-11.5)[b]
G	-9.5	–[a]	-8.3	-6.8
H	-7.1	-6.8	-14.0	-9.6
J	(-6.6)[b]	(-6.7)[b]	-6.2	(-7.1)[b]
K	-2.1	-2.5	0.5	-0.4

[a]Geometry optimization did not converge to a stable structure. [b]Imaginary frequency. The values have only been included for comparison, by neglecting the imaginary frequency in the zero-point energy.

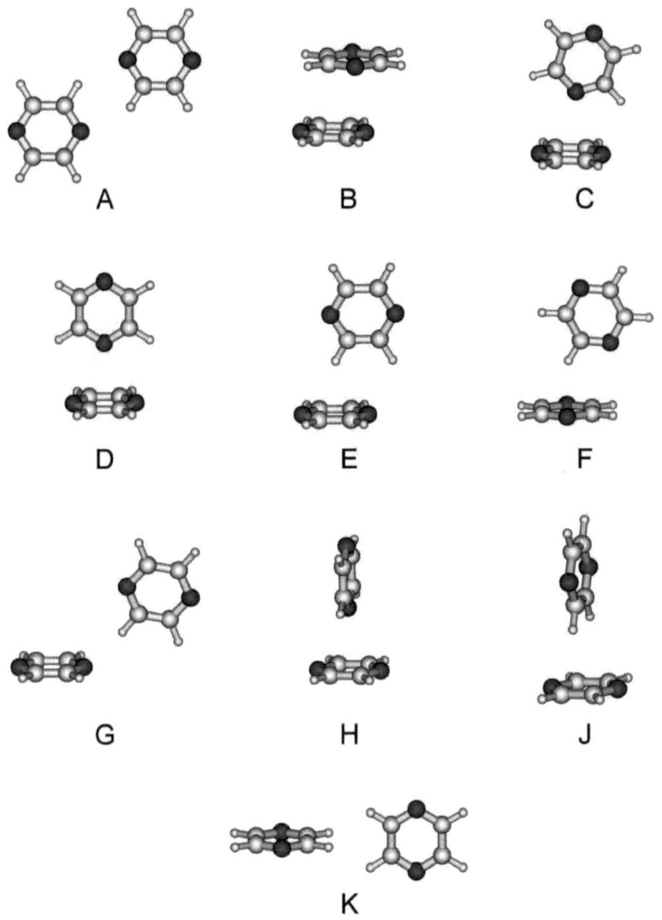

Figure 6.3: *Schematic representation of the calculated pyrazine dimer structures, sorted with increasing B3LYP-D/TZVP energy. The most stable dimer (A) has two stabilizing CH· · · N hydrogen bonds, while the second most stable dimer (B) is the well known stacked, cross-displaced dimer.*

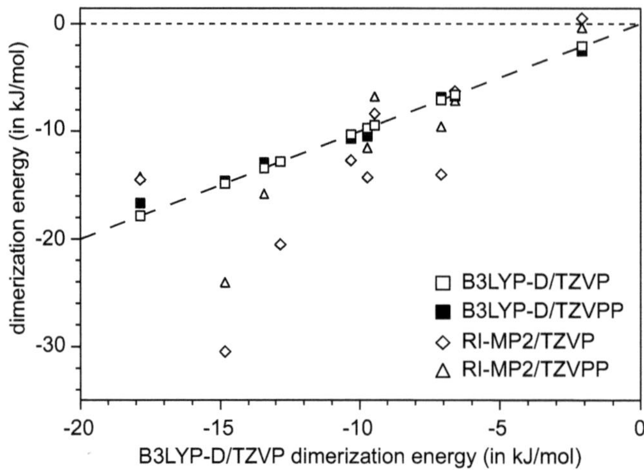

Figure 6.4: *Dimerization energies [E(dimer) − 2E(monomer)] of various pyrazine dimer structures at different levels of theory plotted as a function of the B3LYP-D/TZVP dimerization energy. Both RI-MP2 and B3LYP-D yield similar stabilities for the CH$\cdots$N hydrogen bonded dimer, while MP2 overestimates the stability of the T-shaped and stacked isomers, where dispersion has a larger contribution to the overall stability. See Table 6.1 and Figure 6.3 for a key to the respective dimer structures.*

calculations yield a higher stabilization energy by up to 100 % (see Table 6.1 and Ref. [64]). MP2 and MP4 calculations are known to overestimate the stability of the benzene dimer, [81] whereas DFT-D calculations are again in good agreement with the CCSD(T) benzene dimer energies. [79, 80] This is also illustrated in Figure 6.4 where we plotted the dimerization energies of Table 6.1 as a function of the B3LYP-D/TZVP energies. The latter has been chosen as reference, because it has the largest set of stable local minima of all our calculations. From Figure 6.4 it is apparent that the predicted MP2 stabilities are much larger than the DFT-D values, but only for those isomers which have large contributions of dispersion interactions to the overall stability. On the other hand, both methods yield similar dimerization energies for the planar, doubly-bridged dimer A (Figure 6.3). It is the most stable isomer in our B3LYP-D calculations and is stabilized by two CH$\cdots$N contacts. In this case the agreement between B3LYP-D and RI-MP2 calculations is expected, because both methods provide reasonable descriptions of clusters which are stabilized mainly by hydrogen bonding. Overall B3LYP-D seems to give a reliable description of the potential energy surface of the pyrazine dimer.

In order to facilitate an assignment of the experimental infrared spectra (Figure 6.2) to individual cluster structures, we also calculated harmonic vibrational frequencies for each predicted isomer. The results of the B3LYP-D/TZVP calculations are shown in Figure 6.5 as stick spectra. Also shown are the infrared spectra of isomers 1 and 2. Relative energies (in kJ mol^{-1}) are given in parenthesis. Almost identical spectra were obtained at the B3LYP-D/TZVPP level. MP2 calculations showed only minor differences.

Following our earlier argument, isomer 1 probably involves direct CH$\cdots$N contacts. Among the structures in Figure 6.3 such a direct interaction is only present in isomers A and G. Since dimer G is 8.39 kJ mol^{-1} higher in energy, we assign the spectrum of isomer 1 to the planar, doubly-bridged most stable dimer A. Its stick spectrum is in qualitative agreement with the experiment, though the overlap is not perfect. The origin of the absorption at 3022 cm^{-1} remains unclear. Pyrazine itself has a weak CH stretching absorption at 3018 cm^{-1} [73] that is not discernible at the signal-to-noise level in Figure 6.2 and might gain intensity in the dimer. Another possibility is Fermi resonance between the strong CH stretching vibrations and overtones or combination bands of ring stretching or of CH bending vibrations of the same symmetry. The energetic resonance condition of the Fermi coupling might

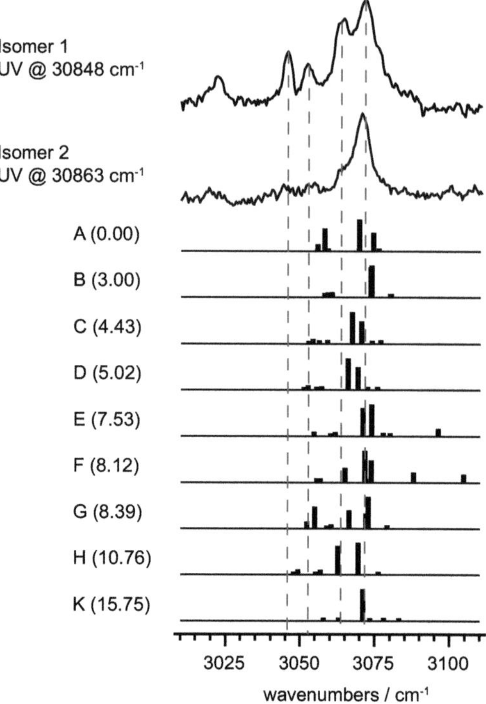

Figure 6.5: *Comparison of the experimentally obtained infrared spectra of isomers 1 and 2 with calculated and scaled stick spectra at the B3LYP-D/TZVP level. Calculated harmonic frequencies were scaled by 0.9694 in order to match the experimental and calculated CH stretching frequency of the pyrazine monomer. Relative dimer energies (in kJ mol^{-1}) at the B3LYP-D/TZVP level are given in parenthesis.*

be favored by the observed frequency shifts in the hydrogen bonded dimer A and explains the absence of a corresponding band in the spectrum of the stacked isomer B. [82]

Isomer 2 shows only a single strong absorption in agreement with the calculated spectra of structures B and K. They are the only isomers which show a single strong IR absorption with all other vibrations having almost zero intensity. We assign the spectrum of Isomer 2 to the stacked, crossed-displaced structure B because of its higher stability. There the CH bonds are only indirectly influenced by the π-stacking and the vibrational frequencies of the individual pyrazine units are almost identical to those of the monomer. The absorption at 3071 $cm^{-1}$of isomer 2 (Figure 6.2) is then an overlap of the bands of the two symmetrically inequivalent pyrazine subunits. The T-shaped structures C and D are only 1.4 and 2.0 kJ mol^{-1} higher in energy than B, but there the CH bonds of the stem are more directly influenced by dimer formation. As a consequence the vibrational frequencies of the top and stem molecules differ and the calculated spectra show two bands of similar intensity $\approx$4 cm^{-1} apart. Such a splitting is well within the resolution of our experiment but is not observed in the infrared spectrum of isomer 2.

6.4 Summary

We presented the IR-UV double resonance spectrum of two pyrazine dimers in supersonic jets. Isomer 1 has a planar structure, stabilized by two CH$\cdots$N contacts, while isomer 2 is the well known cross-displaced, stacked dimer. The assignment is supported by B3LYP-D/TZVP calculations. Until now we have no indication for the presence of a T-shaped structure. Isomers 1 and 2 are the two most stable structures in our B3LYP-D calculations, while MP2 probably tends to overestimate the stability of the stacked and T-shaped structures. Our results are in agreement with previous theoretical work on pyridine dimers. [77] It should be noted, that the most stable doubly-bridged dimer, which is present in supersonic jets, does not directly reflect the structural motif of the pyrazine crystal. [53, 54, 55] The latter is formed by CH$\cdots$N catemer synthons rather than dimers [53], although both are stabilized by CH$\cdots$N contacts. Future experiments are planned to investigate pyrazine trimers and tetramers. These clusters could profit from a combination of stacking and CH...N interactions and could adopt a local configuration similar to the catemer synthons.

Acknowledgement:

This work has been supported by the Deutsche Forschungsgemeinschaft (FOR 618-TPE)

7 Clusters of Pyrazine with Acetylene

Clusters of pyrazine and acetylene are a very interesting model system for the competition of CH...π and CH...N interactions. It is very similar to the benzene-acetylene system we discussed carefully in the sections 8 and 9.

The Lewis formula of pyrazine is presented in figure 7.1.

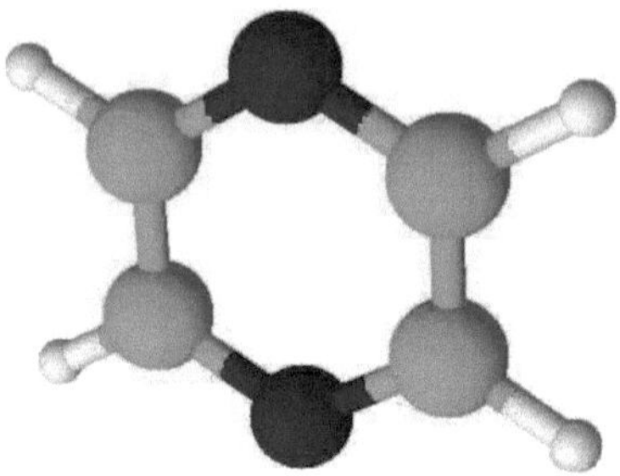

Figure 7.1: *Lewis formula of pyrazine*

The structural assignment of different pyrazine acetylene clusters and their formation probabilities in the molecular beam can unveil, whether CH...π or CH...N interactions dominate the molecule's isomerization under kinetic control.

In order to identify the ideal measurement conditions for this system, the laser energies, temporal position in the gas pulse and the sample temperature were varied. Starting at 80°C, the sample temperature was decreased to room temperature still showing significant ion signal intensity. Figure 7.2 shows time of flight mass spectra recorded on the pyrazine (Pz), pyrazine with one acetylene (PzAc) and pyrazine with two acetylenes ($PzAc_2$) masses at 80°C (top trace), 50°C (middle trace) and room temperature (lower trace). Pyrazine was expanded in pure helium in the top trace, while the middle and lower traces where expanded in a mixture of pyrazine

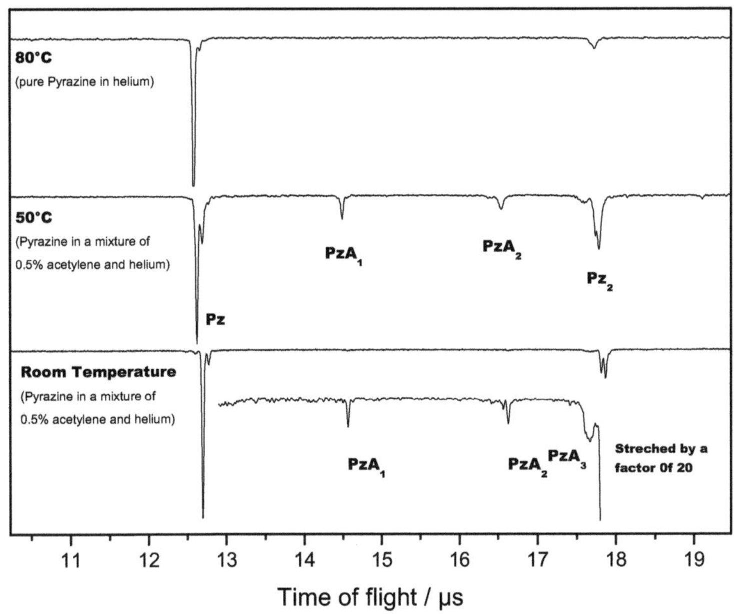

Figure 7.2: *Time of flight mass spectrum of Pyrazine excited at 31400 cm^{-1} and ionized at 193 nm at different source chamber temperatures. In the top trace, Pyrazine is expanded at 80° C and in pure Helium and the two lower traces show expansions at 50° C and room temperature respectively in a mixture of 0.5% acetylene and Helium*

with 2% acetylene and helium. The spectra show the Pz monomer mass, the Pz dimer mass and the masses of the $PzAc_{1\text{-}3}$ clusters.

Especially the ion signals of the cluster masses were weak for room temperature experiments, but the by this expectably low amount of higher clusters and thus decreased fragmentation are ideal for an unperturbed investigation of lower clusters.

Since Pz has a very high ionization potential of 9.3 eV, what corresponds to 74960

cm^{-1}, and the origin is situated at 30876 cm^{-1}, a wavelength of 44100 cm^{-1} is needed to reach the first ionization potential. Hence, the pump laser should be set to 225 nm, which is hardly possible with regular laser dyes (wavelengths below 233 nm are not producible in the standard setup).

Due to the high ionization potential, probing with 193 nm was the only feasible method. REMPI spectra of the Pz monomer and the $PzAc_{1\text{-}2}$ clusters are presented in figure 7.3.

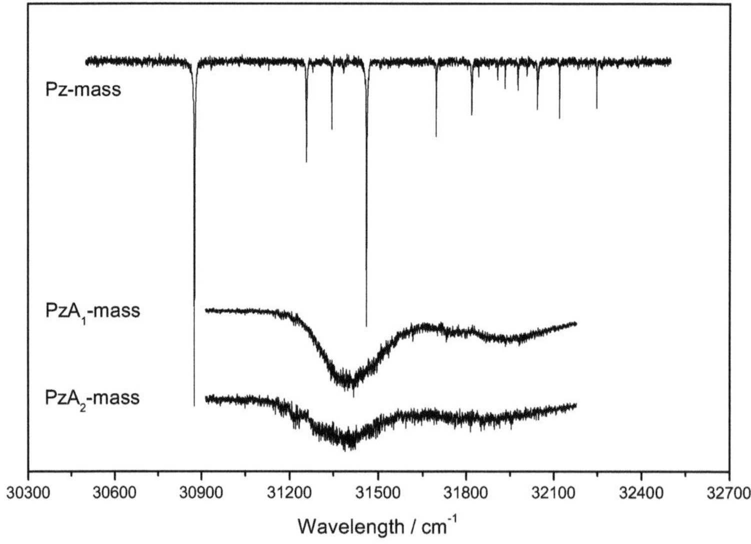

Figure 7.3: *Two color REMPI spectrum of Pz (expanded in pure helium) and it's first two clusters with acetylene. The spectra were recorded at room temperature, while Pz was expanded in a mixture of 0.5% acetylene and helium and ionisation was performed at 193 nm without any delay*

The spectrum on the pyrazine monomer mass (recorded in absence of acetylene) shows several narrow features corresponding to the origin and vibrational transitions at a line width of 1.4 cm^{-1}. Unfortunately, the spectra on the cluster masses just show unstructured and broad electronic spectra, which are exactly the same on each trace. This behavior represents a very strong fragmentation of higher clusters to lower lying fragment masses, as discussed in section 2.2.3. This strong fragmentation will directly result from the approximately 8000 cm^{-1} excess energy, when ionizing at 193 nm.

Reducing the sample temperature to room temperature decreases the molecule's excess energy as well, as it is directly correlated to the final molecular temperature after expansion. Furthermore, the concentration of acetylene molecules in the acetylene co-expansion defines the number of collisions between both types of molecules and hence the number and size of aggregates formed. Figure 7.4 shows IR/UV double resonance spectra on the pyrazine monomer mass recorded at 50°C and a high acetylene concentration of 2% in the top trace and at room temperature and a concentration of 0.5% acetylene in the lower trace.

The upper trace shows two negative bands at 3076 cm^{-1} and 3211 cm^{-1}, which correspond to an increase in ion signal intensity and thus fortified fragmentation by higher clusters. When reducing the number of clusters formed and the sample temperature, a spectrum only showing the CH-stretch vibration typical for pyrazine at 3070 cm^{-1} is visible. This example presents the importance of carefully adapting the experimental conditions to the investigated system.

By further reducing the laser energies (pump = 190 μJ and probe less than 10 μJ) and optimizing the temporal position in the gas pulse, IR/UV double resonance

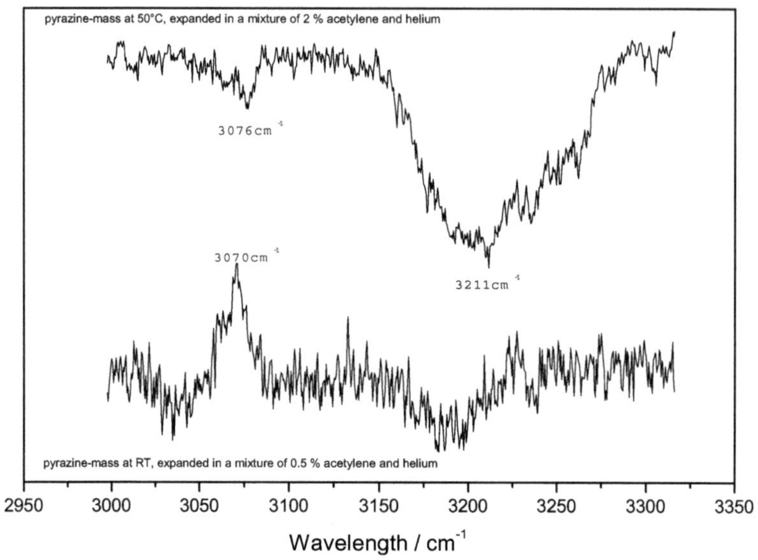

Figure 7.4: *Ir/UV-spectrum recorded on the pyrazine mass showing the strong influence of the acetylene concentration and the sample temperature on the degree of fragmentation. The top trace represents the pyrazine REMPI when expanding in a 2% acetylene helium mixture at 50°C and the top trace was recorded at room temperature and with a 0.5% acetylene helium mixture. The lasers were not delayed and the laser energies were below 10 μJ for 193 nm and 190 μ for the pump laser*

spectra on the Pz monomer and $PzAc_{1-2}$ could be recorded (see figure 7.5).

The IR spectra show very broad bands on all masses, which are still significantly influenced by fragmentation. Since no two color signal was visible for ionization with 233 nm, only a new setup enabling probe wavelengths of 220 to 225 nm can enable the recording of narrow spectra and allow an unambiguous assignment of structures by adequate quantum chemical calculations.

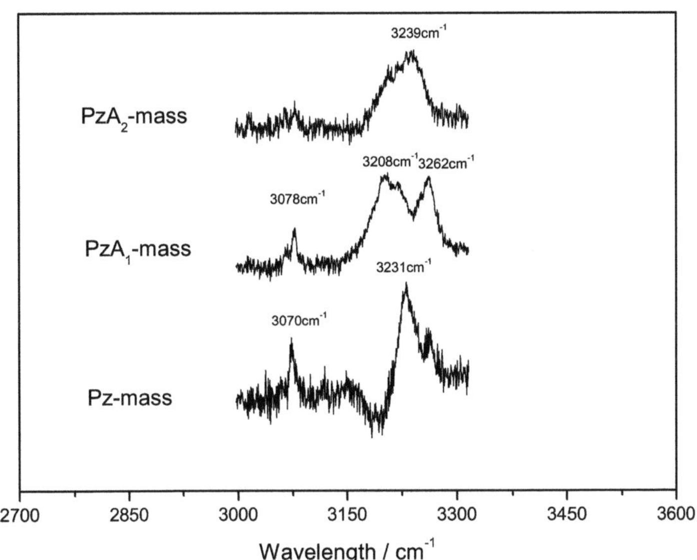

Figure 7.5: *IR/UV double resonance spectra of pyrazine and the first two clusters with acetylene. Spectra were recorded at room temperature expanding in a 0.5 % mixture of acetylene with helium. The lasers were not delayed and the laser energies were below 10 μJ at 193 nm and 190 μJ at the pump wavelength*

8 Cocrystals of Benzene with Acetylene

Isomer Selective Vibrational Spectroscopy of Benzene-Acetylene Aggregates: Comparison with the Structure of the BA Cocrystal

M. Busker, Th. Häber, M. Nispel, K. Kleinermanns

Institut für Physikalische Chemie und Elektrochemie 1, Heinrich-Heine Universität-Düsseldorf, 40225 Düsseldorf, Germany kleinermanns@uni-duesseldorf.de

Published in Angewandte Chemie Int. ed. (2008)

Cocrystals with defined molecular composition can be synthesized by co-condensation of gaseous compounds in fixed molar ratios followed by multiple heating/cooling cycles. Cocrystals with 1:1 and 1:2 ratios and with different structures have been assembled in this way. [83, 84] The basic structural motifs of the unit cell in principle are comparable to nanocrystals (clusters) synthesized in gas jets by adiabatic cooling. However, cooperative effects may lead to isomeric crystal structures which do not necessarily represent the global minimum structures of the clusters. But since supersonic jet cooling is a non-equilibrium process, higher energy isomers are often formed as well. That opens the possibility to directly study isolated unit cell structures of cocrystals in the form of clusters. In the following we report the structures of small benzene-acetylene clusters and compare them to the structure of the 1:1 cocrystal. Strong CH$\cdots\pi$ interactions (typically $\leq$ 2.5 kcal/mol) [85] have found broad interest due to their importance for the stabilization of supramolecular aggregates, crystal packing, molecular recognition and for the folding of proteins. [86, 87, 88] A typical example of such a strong CH$\cdots\pi$ interaction is the T-shaped benzene-acetylene dimer. We decided to investigate benzene-acetylene clusters (BnAm) for direct comparison with the 1:1 cocrystal. In this paper we present infrared spectra of BA2, BA3 and B2A clusters. BA2 forms two isomers in supersonic jets, but only one isomer has been characterized by IR spectroscopy previously. [89, 90] Here we report the IR spectrum of the other isomer for the first time. It has a double T-shaped structure (see below) which is also found in the 1:1 cocrystal along the c-axis (8.1). This isomer might be the seed cluster in crystal growth. Experimentally, IR spectra of the acetylenic CH stretching vibration of benzene/acetylene aggregates have been observed in bulk solution, [91] Ar matrices, [92] and supersonic jets. [89] NMR measurements point to a close CH$\cdots\pi$ contact. [93] The structure of the 1:1 benzene-acetylene cocrystal [83] has a basic packing motif of first neighbours consisting of T-shaped BA arrangements (8.1).

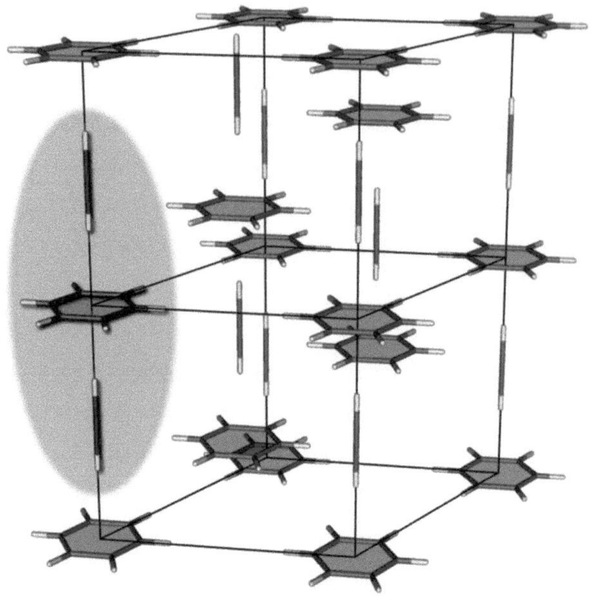

Figure 8.1: *X-ray structure of the benzene/acetylene 1:1 cocrystal [83] indicating the packing motif of T-shaped BA units. Each unit cell is formed of three BA dimers.*

Previous gas phase studies have concentrated on small B1Am clusters (m = 1, 2, 3). Resonant two photon ionisation (R2PI) spectroscopy [90, 89, 94, 95] revealed absorption bands at +137 cm^{-1} (BA), +127 cm^{-1} (BA2, isomer 1), +123 cm^{-1} (BA2, isomer 2) and +116 cm^{-1} (BA3) relative to the 601 band of benzene [96] at 38606 cm^{-1}. The blue shift indicates a reduced cluster stability in the electronically excited state due to a lower π-electron density in the benzene ring upon π^* excitation, as in other π-hydrogen bonded clusters. [97, 98] High level ab initio calculations predict a T-shaped structure of BA. The acetylene lies on the C6 symmetry axis of the benzene ring and forms a π-hydrogen bond with the aromatic π-system. [99, 100, 51, 52] IR-UV double resonance experiments support a T-shaped BA structure, whereas isomer 2 of BA2 forms a ring like structure. [89] Very little is known about the structure of the other benzene-acetylene aggregates which might correlate with the building blocks of the benzene/acetylene cocrystal. Figure 2 shows the IR-UV ion dip spectra of the two isomers of BA2 with the UV laser tuned to 601 + 127 cm^{-1} (isomer 1) and +123 cm^{-1} (isomer 2). Due to fragmentation of the cluster ions the spectra are observable only at the BA mass but velocity map imaging allowed for an unambiguous assignment to BA2. [91] Also shown are the calculated and scaled stick spectra of the most stable cluster structures, sorted with increasing energy from top to bottom. Isomer 2 shows two absorptions at 3259.2 and 3263.9 cm^{-1}. As mentioned above, isomer 2 has been assigned to the ring structure A previously, based on comparison with quantum chemical calculations. [89] The first acetylene forms a C-H$\cdots\pi$ hydrogen bond with benzene, while the second acetylene binds to the π-system of the C≡C bond of the first acetylene and docks sideways to the C-H bonds of the benzene ring. Only structures A and C are predicted to feature two closely spaced absorptions of similar intensity, in good agreement with the experimental spectrum of isomer 2. They differ in a 30° rotation of the benzene ring. The overall spectral pattern is largely independent of the basis set size (8.2).

By contrast isomer 1 has only a single absorption at 3267.1 cm^{-1}, close to that of the T-shaped BA dimer at 3266.7 cm^{-1}. [89] This already points to a highly symmetric double T-shaped structure. The calculated spectra show a single absorption only for the second most stable isomer B (Figure 1), which has indeed a double T-shaped structure. The high symmetry leads to a coupling between the C-H stretch vibrations of the two acetylene units and only one infrared active vibration remains. We rule out isomer D because of its much higher energy and because of the red shift of its

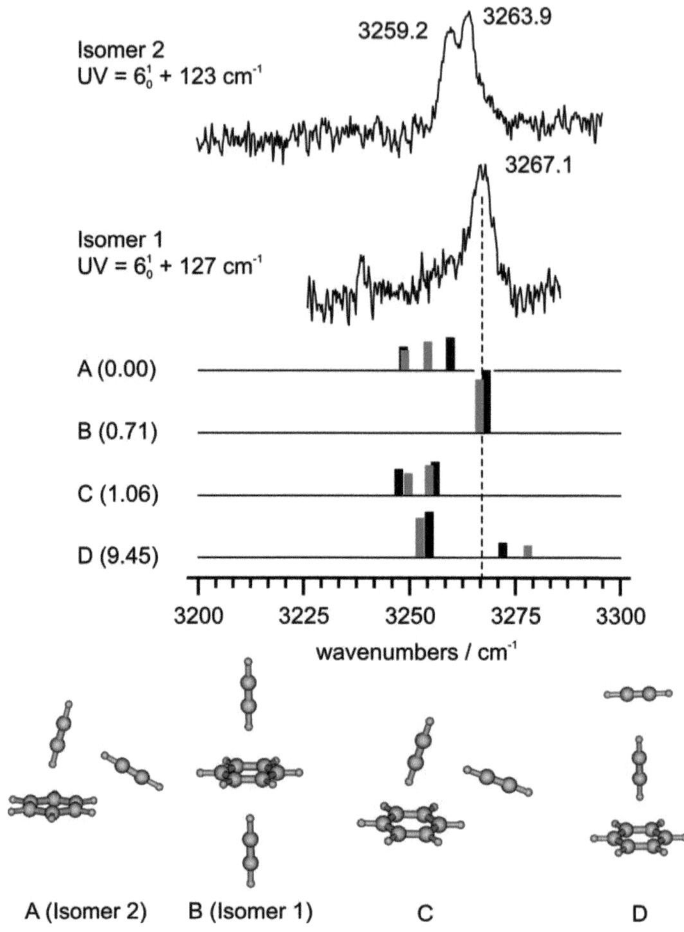

Figure 8.2: *Isomer selective IR-UV ion dip spectra of isomers 1 and 2 of BA2 (top two traces). Also shown are calculated and scaled stick spectra at the RIMP2/TZVP (black) and RIMP2/TZVPP (grey) level. Relative cluster energies (in kJ mol-1) at the RIMP2/TZVPP level are given in parenthesis.*

dominant infrared band relative to the T-shaped BA absorption. The structure of isomer 1 (B) reflects the BA2 motif in the crystal structure (highlighted in Figure 1). The double T-shaped arrangement continues in the BA3 cluster. Its IR-UV spectrum is displayed in Figure 3 and shows three bands at 3252.7, ≈3261 and 3268.6 cm^{-1}. Contrary to BA2 only one isomer has been identified by UV spectroscopy. [90] The absorption at 3268.6 cm^{-1} is almost identical to the T-shaped BA dimer, so that one acetylene molecule is probably in a BA-like arrangement. 8.3 also shows the calculated stick spectra of the most stable BA3 isomers. Considering the agreement between calculated and experimental frequencies, only the spectra of structures A and C match reasonably well with the experiment. They differ only by a 30° rotation of the benzene ring. The vibrational frequencies of B, D, and E are red shifted by about 15 cm^{-1} compared to the experiment. Structure F is ruled out, because it does not predict the observed spectral pattern of three almost equally spaced absorptions of similar intensity. We therefore assign the spectrum of BA3 to the most stable isomer A.

As mentioned above, cluster formation in supersonic jets is not strictly controlled by thermodynamics, but to a large extent by kinetics so that the cluster abundances are influenced by the formation probabilities. If we assume that BA3 is formed by adding acetylene to an already formed BA2 cluster then both isomers of BA2 are precursors to structure A of BA3, whereas structures B and D can only be formed by adding acetylene to isomer 2 of BA2. Therefore the formation of structure A has a higher probability, which supports our spectral assignment. The other way around, observing only structure A supports a stepwise aggregation mechanism in which acetylene is attached to pre-formed BAm clusters, rather than by first forming acetylene clusters (like the cyclic trimer) and attaching them to a benzene ring, which would lead to structure D of BA3 exclusively. We did not observe any isomers of

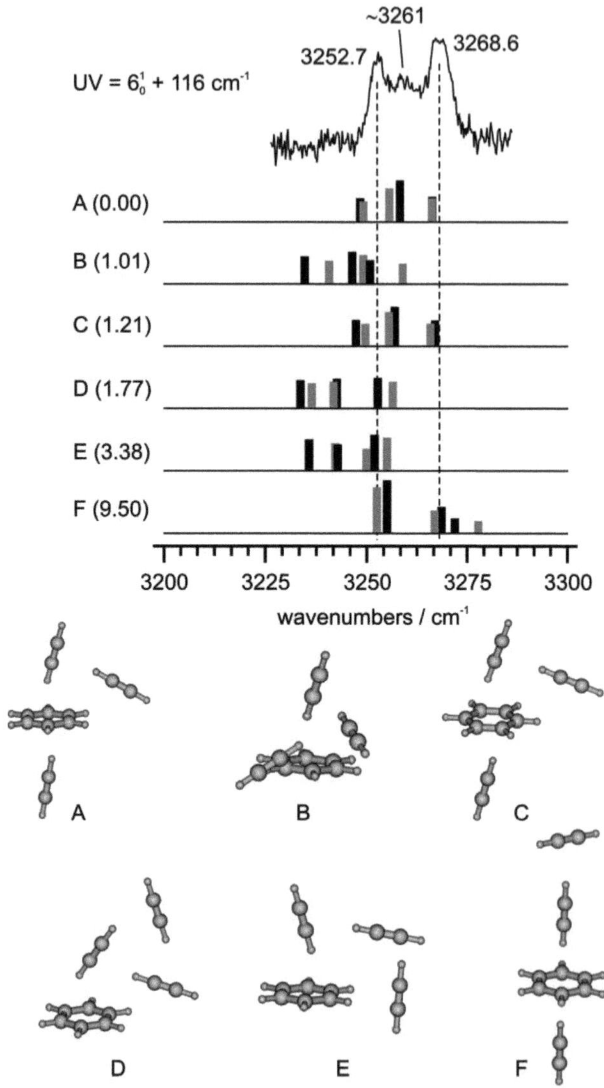

Figure 8.3: *IR-UV ion dip spectrum of BA3 (top trace). Also shown are calculated and scaled stick spectra at the RIMP2/TZVP (black) and RIMP2/TZVPP (grey) level. Relative cluster energies (in kJ mol-1) at the RIMP2/TZVPP level are given in parenthesis.*

BA3 even at higher acetylene concentrations of up to 10%.

Finally 8.4 shows the IR-UV spectrum of B2A obtained on the B2 mass channel and with the UV excitation laser set to 601 + 116 cm^{-1}, together with calculated spectra of the six most stable structures. The experimental spectrum shows only a single absorption at 3259.4 cm^{-1}. Its vibrational frequency is in good agreement with the most stable structure A. We can think of it as the combination of a T-shaped benzene dimer and a T-shaped BA cluster, that is, as a combination of the dominant B2 [81] and BA [89] structures. However, we cannot decide which of the moieties is formed first. We rule out structure B because of its higher energy and its higher vibrational frequency, which is close to that of the T-shaped BA cluster. All other structures are probably too high in energy. In summary, we have presented the isomer selected infrared spectra of BA2, BA3 and B2A. Isomer 1 of BA2 has a double T-shaped structure which is also found in the 1:1 cocrystal along the c-axis. The structure of BA3 points to a stepwise aggregation in which acetylene molecules are successively added to previously formed BAm clusters. As mentioned above, the packing motif of the crystal structure is probably not the most stable structure of clusters of the same size. Therefore it is not unexpected that the most stable BA3 and B2A structures deviate from the arrangement of the molecules in the cocrystal. Investigations of larger clusters are required to determine whether the initial cluster structures undergo isomerization after the attachment of further molecules. The double T-shaped motif, for example, may survive in larger clusters or it may disappear in favour of more compact structures. If it persists then it might be the seed nucleus in the crystallization process of the BA cocrystal. By changing the expansion conditions we are able to facilitate the formation of higher energy isomers like isomer 1 of BA2. Most important, we can differentiate between the various structures by our isomer selective IR/UV double resonance technique,

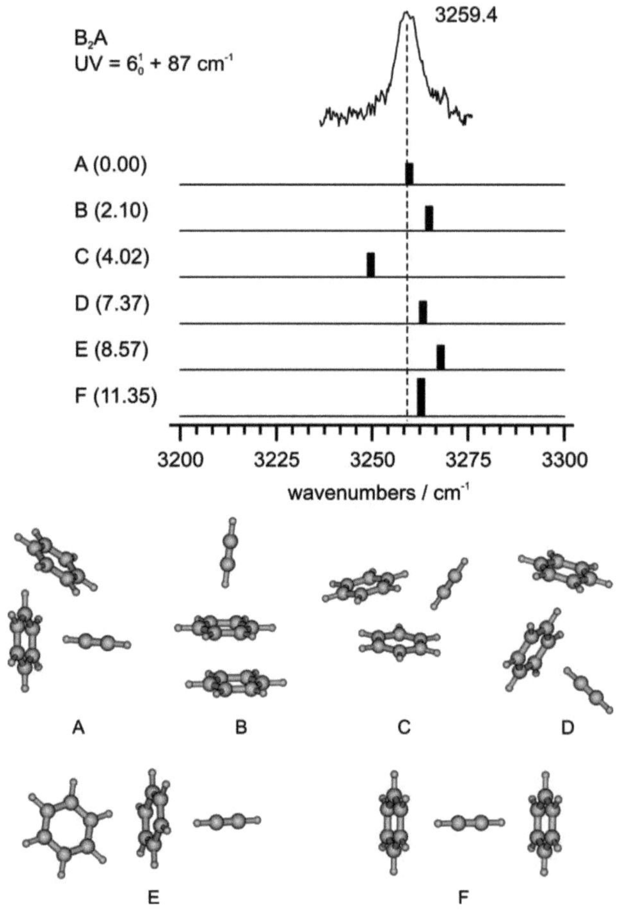

Figure 8.4: *Isomer selective IR-UV ion dip spectrum of B2A. Also shown are calculated and scaled stick spectra at the RIMP2/TZVP level. Relative cluster energies (in kJ mol-1) at the RIMP2/TZVPP level are given in parenthesis.*

assign structures based on quantum chemical calculations and compare them to the structure of the cocrystal.

Methods

The basic principles of our IR-UV experimental setup were described in detail elsewhere. [67, 66, 65] A gas mixture of 0.8% benzene (Acros, >99%), 1.6% acetylene (Air Liquide, 2.6) and 97.6% helium (Air Liquide, 5.0) was expanded through the 300 μm orifice of a pulsed valve (Series 9, General Valve) at a stagnation pressure of 3 bar. The molecules are cooled down to a few Kelvin in the adiabatic expansion and form clusters. The skimmed molecular beam (skimmer diameter 1 mm) crosses the UV excitation (LAS, frequency doubled, $\approx$10 μJ/pulse) and ionization lasers (FL 2002, Lambda Physics, 274 nm, $\approx$120 μJ/J/pulse) at right angle inside the ion extraction region of a linear time-of-flight (TOF) mass spectrometer. A pulsed IR laser beam (burn laser) is aligned collinear to the UV beams and fired 150 ns before the latter. The IR laser frequency is scanned over the vibrational transitions and removes vibrational ground state population if resonant, while the UV excitation laser is kept at a frequency resonant with a vibronic transition of a single cluster isomer. By monitoring the ion mass signal as a function of IR frequency, cluster mass and isomer selective infrared spectra, detected as ion dips can be obtained. IR laser light between 2800 and 4000 cm^{-1} is generated by difference frequency generation in a LiNbO3 crystal and amplified by optical parametric amplification. [68] The ro-vibrational transitions of the NH stretching vibrations of ammonia were used for frequency calibration. The RI-MP2 calculations were performed with the TURBOMOLE V5.9 program package. [41] Calculated harmonic vibrational frequencies were scaled by 0.9351 (TZVP basis set) or 0.9516 (TZVPP) to match the experimental absorption maximum of the asymmetric acetylene C-H stretching vibration of T-shaped BA [89]. All relative energies reported here have been corrected for the zero-point energy (ZPE).

9 Higher Aggregates of Benzene and Acetylene

Towards a Spectroscopical and Theoretical Identification of the Isolated Building-Blocks of the Benzene-Acetylene Cocrystal

Matthias Busker, Thomas Häber, Michael Nispel, Karl Kleinermanns*

Institut für Physikalische Chemie und Elektrochemie 1, Heinrich-Heine Universität-Düsseldorf, 40225 Düsseldorf, Germany kleinermanns@uni-duesseldorf.de

to be submitted

Abstract

We present isomer and mass-selective UV and IR-UV double resonance spectra of the B_2A and B_2A_2 benzene-acetylene clusters. Cluster structures are assigned by comparison with the UV and infrared spectra of benzene, benzene dimer as well as BA_n (n=1, 2, 3) clusters. RI-MP2/TZVPP calculations support our assignment. T-shaped geometries are the dominant structural motifs. For every cluster size the most stable structure was found and identified in supersonic expansions. The observed cluster structures are correlated with possible cluster formation pathways and their role as crystallization seeds is discussed.

Keywords: IR Spectroscopy, Molecular Beams, Clusters, Crystal Growth, Ab Initio Calculations

9.1 Introduction

Cocrystals with defined molecular composition can be synthesized by co-condensation of gaseous compounds in fixed molar ratios followed by multiple heating/cooling cycles, as long as the cocrystal is more stable than the crystals of the pure components. The basic structural motifs of the unit cell in principle are comparable to nanocrystals (clusters) synthesized in gas jets by adiabatic cooling. However, cooperative effects may lead to crystal structures which do not necessarily represent the most stable cluster isomer in the gas phase. But since supersonic jet cooling is a non-equilibrium process, higher energy cluster isomers are often formed along with the most stable forms. That opens the possibility to directly study isolated unit cell structures of crystals or their building blocks in the form of clusters. Isomer and mass selective high resolution spectroscopy can be used to determine the electronic and structural properties of the clusters, which can be used to identify the sequence of cluster formation steps and to relate it to the crystallization process.

Cocrystals of benzene (B) and acetylene (A) with defined molecular composition have been synthesized by co-condensation of gaseous compounds in fixed molar ratios and characterized by x-ray structure analysis. [83] The structure of the 1:1 benzene-acetylene cocrystal has a basic packing motif of first neighbors consisting of T-shaped BA arrangements, stabilized by $\mathrm{CH}\cdots\pi$ interactions between the acidic CH group of acetylene and the π system of the benzene ring. The relatively strong $\mathrm{CH}\cdots\pi$ interaction of typically $\leq 2.5\,\mathrm{kcal\,mol^{-1}}$ energy [85] has found broad interest due to its importance for the stabilization of supramolecular aggregates, crystal packing, molecular recognition and folding of proteins [86, 87, 88]. The benzene-acetylene dimer (BA) is a typical case of such a stronger $\mathrm{CH}\cdots\pi$ interaction. Larger aggregates, for example, B_2A and B_2A_2 are able to form $\pi\cdots\pi$ bonds between the two benzene moieties which might compete with the benzene-acetylene $\mathrm{CH}\cdots\pi$ interaction.

Benzene-acetylene aggregates have already been studied by several groups. IR spectra of the acetylenic C–H stretching vibration have been observed in bulk solution [91], Ar matrices [92], and supersonic jets. [89] NMR measurements show a characteristic high field chemical shift of the acetylenic proton which points to a close CH$\cdots\pi$ contact. v Resonant two photon ionization (R2PI) [90,94,95] and laser induced fluorescence (LIF) spectroscopy [89] revealed UV absorptions at +137 cm^{-1} (BA), +127 cm^{-1} (BA_2, isomer 1), +123 cm^{-1} (BA_2, isomer 2) and +116 cm^{-1} (BA_3) relative to the 6^0_1 band of benzene at 38606 cm^{-1} [101, 96], but only velocity map imaging allowed for an unambiguous assignment of the cluster size. [90]

High level ab initio calculations predict a T-shaped structure of BA. The acetylene is situated on the C_6 symmetry axis of the benzene ring and forms a CH$\cdots\pi$ bond with the aromatic π-system. [99, 100, 51, 52] IR-UV double resonance studies confirm the T-shaped structure of BA and a ring structure for isomer 2 of BA_2. The acetylenic CH stretch vibration is red shifted by −22 cm^{-1} (BA) and −27/−30 cm^{-1} (BA_2, isomer 2) relative to the ν_3 mode of bare acetylene. [89] A comparative study of several aromatics with acetylene showed a positive correlation of the IR red shifts with the π-electron density but only a weak negative correlation with the polarizability of the aromatic ring, indicating that the 1:1 cluster is primarily stabilized by a CH$\cdots\pi$ (hydrogen) bond rather than by van der Waals interaction. [89]

Previously we reported isomer selected IR/UV double resonance spectra of both isomers of BA_2 as well as of BA_3 and B_2A. [102] Isomer 1 of BA_2 has a double T-shaped structure which is also found in the benzene-acetylene 1:1 cocrystal [83] along the c-axis and might be the seed nucleus in the crystallization process of the cocrystal. The structure of BA_3 derives from the double T-shaped BA_2 structure by adding a third acetylene which points to a stepwise aggregation in which acetylene molecules are successively added to previously formed BA_n clusters. Finally, B_2A is a combination of the most stable T-shaped B_2 [57,58,56,59,61,60] and BA [89,51,100,52,99] aggregates. Overall, T-shaped structures dominate the observed benzene/acetylene clusters in supersonic jets. [89, 102]

In order to follow the pathway of crystallization from the simplest seed up to the packing motif of a crystal and to learn more about the interplay between $CH \cdots \pi$ bonds and $\pi \cdots \pi$ interactions we decided to investigate benzene-acetylene clusters containing two benzenes. Here we present a detailed vibrational analysis of the different benzene-acetylene clusters and report the R2PI and IR/UV double resonance spectra of B_2A and B_2A_2. The vibrational spectra will be compared to ab initio calculations for structural assignments. We then relate the cluster structures to its smaller precursors and derive possible pathways for their formation in supersonic jets.

9.2 Experiment

The basic principles of our IR-UV experimental setup were described in detail elsewhere. [65, 66, 67] A gas mixture of 0.8 % benzene (Acros Organics, > 99 %), 1.6 % acetylene (Air Liquide, 2.6) and 97.6 % helium (Air Liquide, 5.0) was expanded through the 300 μm orifice of a pulsed valve (Series 9, General Valve) at a stagnation pressure of 3 bar. The molecules are cooled down to a few Kelvin in the adiabatic expansion where they form clusters.

The skimmed molecular beam (skimmer diameter 1 mm) crosses the UV excitation laser (LAS, frequency doubled, 10 μJ/pulse) and the ionization laser (FL 2002, Lambda Physics, frequency doubled, 120 μJ/pulse) at right angle inside the ion extraction region of a linear time-of-flight (TOF) mass spectrometer in Wiley-McLaren configuration. The UV excitation and ionization lasers are spatially and temporarily overlapped. Resonant two photon ionization (R2PI) spectra are recorded by scanning the frequency of the excitation laser between 38500 and 38800 cm^{-1}, while the ionization laser is kept at a fixed wavelength of 274 nm ($\approx$36500 cm^{-1}). The two-color scheme minimizes excess energy in the ions and reduces the amount of fragmentation after ionization.

For the IR/UV double resonance experiments a pulsed IR laser beam (burn laser, 0.2 cm^{-1} resolution) is aligned collinear to the UV excitation beam (probe laser) and fired 150 ns before the latter. The IR laser frequency is scanned over the vibrational transitions and removes vibrational ground state population if resonant, while the UV excitation laser is kept at a frequency resonant with a vibronic transition of a single cluster isomer. By monitoring the ion mass signal as a function of IR frequency, cluster mass and isomer selective infrared spectra, detected as ion dips, can be obtained. IR laser light is generated by a two-stage setup [68], producing infrared radiation between 2800 and 4000 cm^{-1}(10 Hz, 4 mJ/pulse). The rovibrational transitions of the NH stretching vibrations of ammonia were used for frequency calibration.

The RI-MP2/TZVPP calculations were performed with the TURBOMOLE V5.9 program package. [41] Calculated harmonic vibrational frequencies were scaled by 0.9351 (TZVP) and 0.9516 (TZVPP) to match the experimental absorption maximum of the asymmetric acetylene C-H stretching vibration of T-shaped BA at 3266.7 cm^{-1} [89]. A total number of 900 cluster structures were generated randomly using the GRANADA program [103] and pre-optimized at the PM6 [104] level using the MOPAC 2007 program package. [105] The most stable structures as well as some manually generated structures were taken and further optimized at the RIMP2 level. The structural predictions/diversity of the statistical GRANADA/MOPAC approach were tested beforehand on smaller BA clusters (BA_2, BA_3) where it reproduced all major structures reported previously.

9.3 Results

9.3.1 R2PI Spectra

UV spectra of benzene-acetylene clusters with one benzene molecule (BA_n) have been reported previously [90, 94, 95, 89] and assigned to different cluster sizes by velocity map imaging spectroscopy. [90] To facilitate an easier comparison with the B_2A_n spectra and to illustrate similarities we present here a comprehensive summary of the UV spectra of all benzene-acetylene clusters measured so far. Figure 9.1 shows the two-color resonance enhanced two-photon ionization (R2PI) spectra of BA_n (n =1, 2, 3) and B_2A_n clusters (n = 1, 2) in the region of the 6^1_0 band of benzene, together with the benzene monomer (B) [101, 96] and benzene dimer (B_2) [56, 57, 61]. Arrows indicate relative frequency shifts. Cluster size assignments are given by labels.

Benzene-acetylene clusters undergo extensive fragmentation after ionization by loosing one acetylene, so that the cluster spectra can only be obtained on the B_xA_{n-1} mass channels, even under the gentle conditions of our two-color experiment (ionization at 274 nm). Only the spectrum of BA was observed on the B mass channel as well as on the BA mass channel. Within the accuracy of our experiment the relative shifts of the BA_n cluster bands are in agreement with the literature [94, 89, 90, 95] to within 1 cm^{-1}, except for isomer 1 of BA_2, which is off by 3 cm^{-1}. The reason for this discrepancy is still unclear.s

The simplest benzene-acetylene cluster, BA, has a T-shaped structure. [89, 51, 100, 52, 99]. Its UV absorption is blue shifted by 138 cm^{-1} relative to the 6^1_0 band of the benzene monomer. Apparently the cluster stability is reduced in the electronically excited state due to a lower π-electron density in the benzene ring upon $\pi\pi^*$ ex-

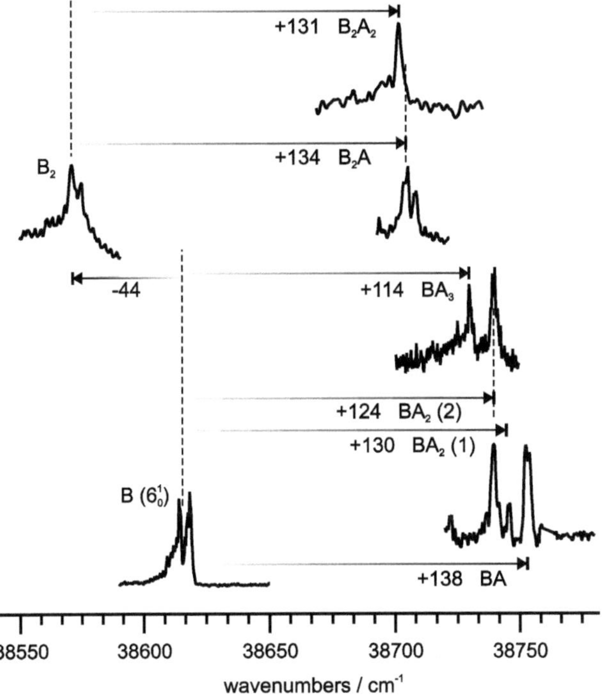

Figure 9.1: *Two color R2PI spectra of benzene (B)-acetylene (A) clusters (BA_n and B_2A_n) in the region of the 6^1_0 of benzene at 38606 cm^{-1} [101, 96]. Also shown is the R2PI spectrum of the benzene dimer (B_2) [56, 57, 61] and the relative shifts of the cluster absorptions with respect to the B or B2 bands. (Ionization at 274 nm)*

citation, as in other π-hydrogen bonded clusters. [97, 98] Successively adding more acetylene molecules leads to a small red-shift of the vibronic transitions.

The benzene dimer has a T-shaped structure too but shows an absorption 44 cm^{-1} to the red of the benzene monomer [57, 56, 61], which is split due to a coupling to a librational mode. [61] The excitation is mainly localized in the "stem" benzene moiety, while the "top" moiety shows long vibrational progressions with maximum absorptions less than 10 % of the "stem". [57, 58, 59, 60] In other words, the π-system of the absorbing chromophore is not part of a CH$\cdots\pi$ bond as it is in BA. This also explains the different frequency shifts (red vs. blue) of B_2 and BA.

The UV absorption of B_2A is blue shifted by 134 cm^{-1} relative to B_2. Following the above argument we can immediately conclude that (1) the acetylene molecule forms a CH$\cdots\pi$ bond with the absorbing chromophore in B_2A and (2) the structure of B_2A probably derives directly from B_2 by adding acetylene in a T-shaped arrangement, because the magnitude of the blue-shift is almost the same as between B and BA. This is supported by IR-UV double resonance spectra in comparison to ab initio calculations (see below and Ref. [102])). The nature of the weak absorption on the blue side of the B_2A band is still unclear. The spacing between the two bands is the same as the splitting of the benzene dimer band, so that the coupling to a librational mode might persist in the B_2A cluster. A second isomer is ruled out as the origin of the weaker band, because the IR-UV spectra of both bands are identical to within less than 1 cm^{-1}.

Finally, B_2A_2 is red shifted by only 3 cm^{-1} relative to B_2A. Therefore the nearest environment around the absorbing chromophore is probably the same as in B_2A. Contrary to B_2A, we did not observe any splitting or other bands on the corresponding mass channel.

9.3.2 IR-UV Spectra

Before discussing the IR-UV double resonance spectrum of B_2A_2, we will summarize our previous measurements which is necessary for an unambiguous structural assignment of the larger clusters. Figure 9.2 shows the IR-UV double resonance spectra of both isomers of BA_2 as well as of BA_3 and B_2A in the region of the ν_3 band of acetylene. [102] The clusters were excited at their respective UV transitions (Figure 9.1) and ionized at 274 nm. Also shown are the cluster structures. Vibrational assignments are color coded. Structural and vibrational assignments are based on comparison with MP2/TZVP and MP2/TZVPP calculations. [102] Not shown here is the spectrum and structure of the T-shaped BA cluster, which absorbs at 3266.7 cm^{-1}. [89]

All cluster structures found so far in supersonic jets are also the most stable isomers at the MP2/TZVPP level of theory. In the case of BA_2 two isomers were identified. Isomer 2 is energetically more stable than isomer 1 by 0.71 kJ mol^{-1}, which is reflected by the different UV intensities (Figure 9.1).

The acetylene molecules can be divided into three groups according to their local environment which is reflected in their vibrational frequencies. Isomer 1 of BA_2 has a double T-shaped structure with one acetylene on either side of the benzene ring. Due to its high symmetry only one vibration has non-zero intensity. Its absorption maximum is almost identical to that of BA, since the T-shaped local environment of the acetylene molecules is the same in both clusters. Isomer 2 of BA_2 has a ring-like structure and shows two absorptions to the red of BA. The higher absorption frequency of 3263.9 cm^{-1} (blue) is almost exclusively localized on the very acetylene molecule, which forms the $CH\cdots\pi$ bond with the benzene ring. The absorption at 3259.2 cm^{-1} (green) is localized on the other acetylene which forms a CH hydrogen bond with the C≡C triple bond of the first acetylene and "docks" side-ways between two CH bonds of the benzene ring.

BA_3 derives from isomer 2 of BA_2 by adding a third acetylene molecule underneath the benzene ring in a T-shaped geometry so that it forms another $CH\cdots\pi$ bond (Figure 9.2). The vibrational frequency of the latter (red) is again similar to that

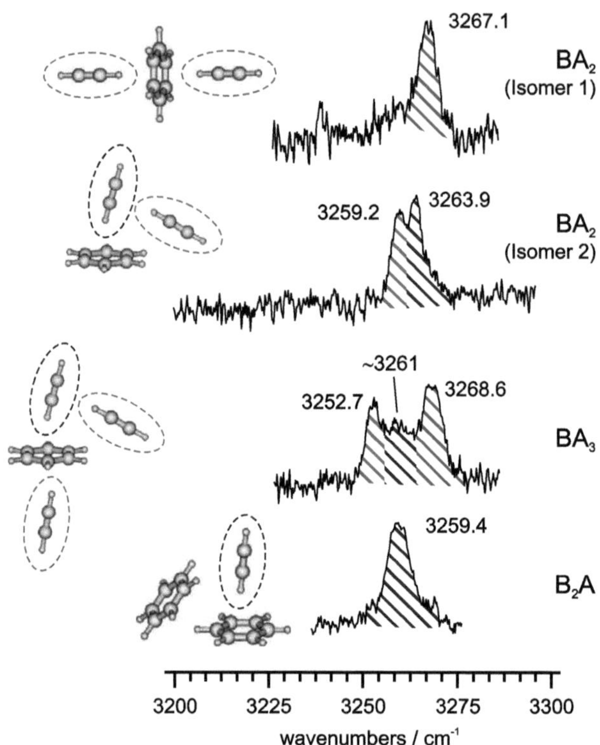

Figure 9.2: *IR-UV double resonance spectra of both isomers of BA_2, of BA_3 and of B_2A. [102] The clusters were ionized at their respective UV transitions (see Figure 9.1) and ionized at 274 nm. Shown to the left are the corresponding cluster structures. Vibrational assignments are color coded. The structural and vibrational assignments are based on comparison with MP2/TZVPP calculations. [102]*

of BA, because there is no other acetylene attached to it. The relative order of the vibrational frequencies of the two other acetylene molecules remains unchanged, but they are slightly red-shifted compared to BA_2 (isomer 2).

Finally B_2A shows a single absorption at $3259.4\,cm^{-1}$. From the R2PI spectra we concluded that acetylene forms a $CH\cdots\pi$ bond with the absorbing chromophore, that is, with the "stem" of the T-shaped arrangement of the two benzene molecules. This is reflected by the most stable cluster isomer depicted in Figure 9.2. [102] It is interesting to note, that the acetylene molecule in B_2A is in a similar arrangement as the first acetylene in BA_2 (isomer 2), but now the second acetylene is replaced by another benzene ring. Apparently the interaction with the "top" benzene molecule in B_2A leads to a slightly stronger red-shift than the interaction with the second acetylene in BA_2. In summary, both the formation of a $CH\cdots\pi$ bond and the interaction with the C≡C triple bond lead to a red-shift of the asymmetric C–H stretching vibration.

Figure 9.3 shows the IR-UV double resonance spectrum of B_2A_2 with the UV excitation laser tuned to $6^1_0 + 87\,cm^{-1}$ and an ionization wavelength of 274 nm. The spectrum shows two absorptions of similar intensity at 3239.5 and $3257.7\,cm^{-1}$. Also shown are the calculated and scaled stick spectra of different B_2A_2 isomers with a relative energy ΔE of less than $9\,kJ\,mol^{-1}$ compared to the most stable isomer A. A total number of 22 isomers were preoptimized at the RIMP2/TZVP level. Some of them converged to the most stable structures shown in Figure 9.4. For the most stable isomers, geometries were further optimized and harmonic frequencies calculated also at the RIMP2/TZVPP level. Contrary to BA_2 and BA_3 [102] the relative energies of some isomers changed dramatically between using the TZVP and TZVPP basis sets, although structural changes are small. Among the most stable isomers

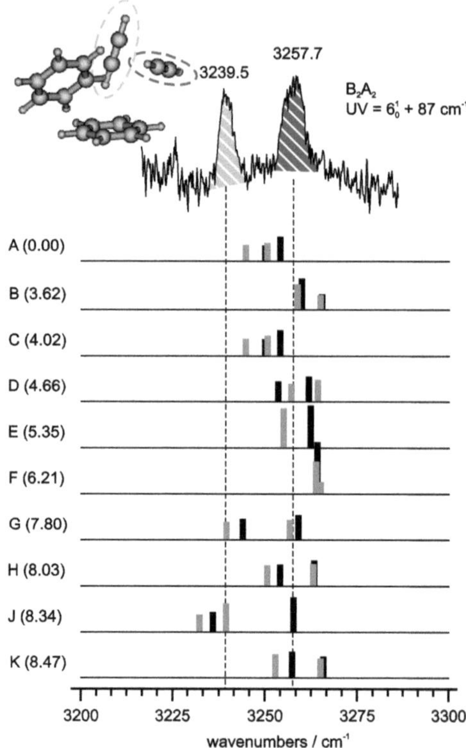

Figure 9.3: *IR-UV double resonance spectrum of B_2A_2 obtained with the UV excitation laser tuned to 38693 cm^{-1}. Also shown are the calculated and scaled stick spectra of selected isomers ($\Delta E < 9$ kJ mol^{-1}) at the RIMP2/TZVP (black) and RIMP2/TZVPP (grey) levels of theory. Relative cluster energies (in kJ mol^{-1}, including ZPE) at the RIMP2/TZVPP level are given in parenthesis. The most likely structure that we assign to the spectrum is depicted in the upper left corner. Vibrational assignments are color coded. See Figure 9.4 for a complete list of the isomeric B_2A_2 structures.*

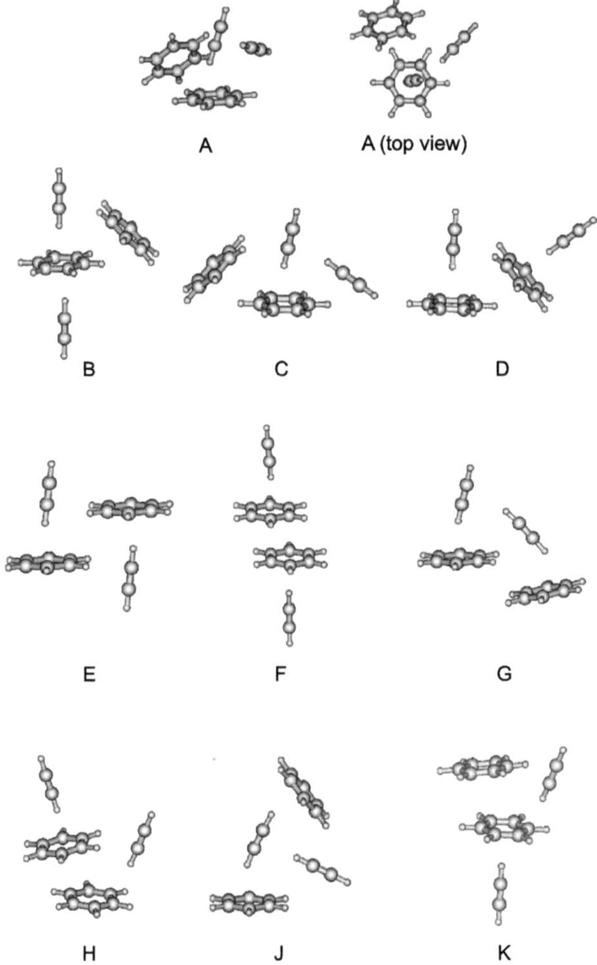

Figure 9.4: *Optimized structures of selected isomers ($\Delta E < 9\,kJ\ mol^{-1}$) of B_2A_2 at the RIMP2/TZVPP level of theory.*

in Figure 9.4 such a dramatic effect was only observed for isomer E, which is the most stable isomer on the TZVP level, followed by isomers B, D and C with relative energies of 0.85, 0.91 and 1.11 kJ mol^{-1}, respectively, compared to E. Especially the side-ways interaction of the C≡C triple bond with the CH bonds of the benzene ring (like in isomer E) seems to depend strongly on the basis set. Higher level ab initio calculations might be needed to describe the subtle interactions present in this weakly bound system more precisely.

If we just look at the vibrational frequencies and compare them to those of the smaller clusters in Figure 9.2, we might imagine that the band at 3257.7 cm^{-1} belongs to an acetylene which forms a $CH\cdots\pi$ hydrogen bond, either to a benzene ring or a C≡C triple bond, while at the same time its own C≡C triple bond acts as an hydrogen acceptor (blue and green bands in Figure 9.2). However, none of our previous spectra showed a red-shift as large as the second band of B_2A_2 at 3239.5 cm^{-1}. To explain such a large red-shift the second acetylene either forms two $CH\cdots\pi$ bonds or its triple bond acts as a hydrogen acceptor for two molecules in addition to the $CH\cdots\pi$ bond. Of the most stable structures in Figure 9.4 only isomers A, C, G and J fulfill that condition which is confirmed by the calculated vibrational frequencies in Figure 9.3. From the UV spectrum (Figure 9.1) we know that B_2A_2 probably has the same T-shaped-like arrangement of the two benzene rings as B_2A. That leaves us with isomers A and C as the likeliest B_2A_2 structures in our experiment. Therefore we tentatively assign the B_2A_2 spectrum in Figure 9.3 to the most stable isomer A. Its structure is also shown in the upper left corner of Figure 9.3 together with the vibrational assignment.

9.4 Discussion

Based on the observed structures we can now try to unravel the possible cluster formation pathways (Figure 9.5). Starting from the simplest benzene-acetylene cluster, BA, both isomers of BA_2 are formed by adding another acetylene. The ring-like isomer 2 is the most stable structure and is formed with larger abundance than isomer 1. This is confirmed by its more intense UV absorption, if we assume similar UV transition probablilities for both isomers. As mentioned earlier, cluster formation in supersonic jets is not strictly controlled by thermodynamics, but to a large extent by kinetics so that the cluster abundances are influenced by the formation probabilities as well. If we assume that BA_3 is formed by adding acetylene to an already formed BA_2 cluster then both isomers of BA_2 are precursors to the observed structure of BA_3, which is the most stable BA_3 isomer at the MP2/TZVPP level. [102] Therefore its isomeric structure is favored by both, thermodynamics and kinetics. No other isomers of BA_3 have been found so far, even at higher acetylene concentrations of up to 10 %. Stated the other way around, observing only the BA_3 structure in Figure 9.5 supports a stepwise aggregation mechanism in which acetylene is attached to pre-formed BA_n clusters, rather than by first forming acetylene clusters (like the T-shaped A_2 dimer or cyclic A_3 trimer) and attaching them to a benzene ring.

We can derive a similar scheme for the B_2A_n clusters, but now each cluster in principle has two precursors. B_2A might be formed from B_2 by adding acetylene or from BA by adding benzene. Similarly B_2A_2 has B_2A and BA_2 as possible precursors. So far our experimental data is not sufficient to decide if one of the pathways dominates over the other. Thus, the question remains open whether B_2A_n clusters build upon B_2, BA or both. However, since we have an excess of acetylene in our supersonic expansions, BA is statistically favored over B_2 as precursor.

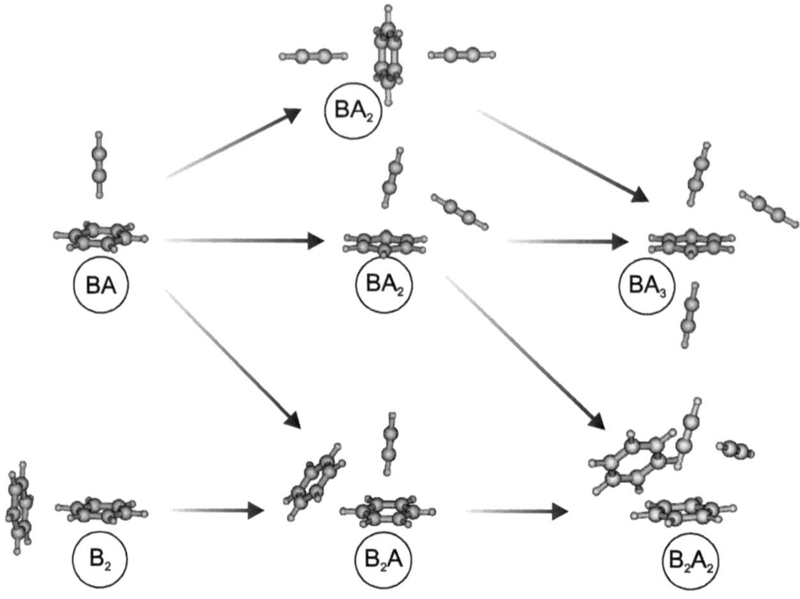

Figure 9.5: *Schematic representation of the benzene-acetylene cluster formation pathways based on the structures that were observed in supersonic jets (this work and Ref. [102]).*

It should be noted that all of the benzene-acetylene clusters found in supersonic jets are not only statistically favored but also thermodynamically the most stable.

We would like to mention in this context that isomer E, which directly resembles part of the unit cell of the benzene-acetylene cocrystal [83], has no direct precursor in Figure 9.5, but requires a rearrangement of the molecules in order to be formed out of the observed structures. A possible precursor/intermediate for such a rearrangement is isomer J. It derives directly from the most abundant BA_2 structure. Shifting the top benzene molecule to the right might result directly in isomer E. Another possible precursor for the T-shaped motifs of the crystal structure is isomer H, which, after rearrangement could yield again a cluster build of two T-shaped BA units. Future experimental and theoretical investigations are planned to search for other isomers, to get a more precise description of the energetics and to explore possible pathways that would lead to structural motifs of the cocrystal.

Acknowledgement:

This work has been supported by the Deutsche Forschungsgemeinschaft (FOR 618-TPE).

10 Summary

During this thesis, REMPI and IR/UV double resonance spectra of 1-methylthymine, small clusters of 1-methylthymine with water, the pyrazine dimers, clusters of pyrazine with acetylene and a large range of clusters of benzene with acetylene were recorded. Structural assignment of the above named species was done by comparison to quantum chemical calculations. Furthermore, several approaches to unveil the nature of the spectroscopically dark state in 1-methylthymine were carried out incorporating oxygen quenching and the characterization of triplet states in the gaseous phase.

The summary is separated into intramolecular interactions, which concern the intrinsic photostability of the DNA base 1-methylthymine (first part) and intermolecular CH...π and CH...N interactions, which determine the molecular structure and by this photochemical properties.

Spectroscopically dark states in biomolecules

We presented comprehensive REMPI and IR/UV spectra of 1-methylthymine and it's first two clusters with water. Structural assignment of the water clusters was possible by comparison to quantum chemical calculations on the RIMP2/ccpVDZ level of theory.

Delayed ionization spectroscopy allowed us to probe the dark intermediate state, which might be responsible for CPD. We found, that the dark state is not quenched in presence of water and that it's lifetime even increases when forming aggregates with water. This result shows, that the photostability of 1-methylthymine is an intrinsic property of the molecule and not based on water cluster formation.

In order to identify the nature of the dark state, we co-expanded with oxygen. The most probable origin of the dark state besides a $^1n\pi^*$ is a $^3\pi\pi^*$ state, which should be quenched in the presence of oxygen.

Since the lifetime was not affected by oxygen co-expansion and a lifetime of 227 $\pm$ 30 ns is typically too short for a triplet state, we suggested the dark state we observed in the gas phase to be the $^1n\pi^*$ state. This statement is supported by the fluorescence lifetime measurements on 1,3 DMU carried out by Kong et. al. [33].

In a last approach we tried to directly measure triplet triplet absorption spectra in the gas phase in 1-methylthymine. We did not see any bands either in the range of the $^3\pi\pi^*$ nor $^1n\pi^*$ transition. Unfortunately, the calculated oscillator strengths are very small for these transitions such that the detection threshold of our setup might impede this approach.

Model systems showing CH...π and CH...N interactions

We identified the structure and electronic resonances of the $B_1A_{1\text{-}3}$ and $B_2A_{1\text{-}2}$ aggregates by comparison to calculations on the RIMP2/TZVP and RIMP2/TZVPP level of theory.

A kinetically controlled cluster formation pathway is described, which results in the statistically and thermodynamically favored structures. Furthermore, an Isomer representing the BA 1:1 co-crystal's basic packing motif is found in our molecular beam experiments, which could be a probable seed crystal in the crystal formation process.

We recorded IR-spectra of the two dimers of pyrazine and structurally assigned them by comparison to quantum chemical calculations. The calculations were performed on the B3LYP-D/TZVP level of theory, which incorporates dispersive interactions.

The dimer structures found show a planar structure, which is stabilized by two CH..N interactions and a stacked cross-displaced structure.

The dimer does not resemble the pyrazine crystal packing motif, since the crystal is made up of CH...N catermer synthons and not dimers. Pyrazine trimers and tetramers could bridge the gap to the crystal structure.

REMPI and IR/UV double resonance spectra were recorded on the masses of the first two clusters of pyrazine with acetylene ($Pz_1A_{1\text{-}2}$). The spectra are unstructured and broad due to a high level of fragmentation. This effect can be reduced when ionizing at 220 nm for example, to reduce the excess energy in the ions and thus increase the barrier to fragmentation compared to ionization at 193 nm.

Outlook

The above stated results show that IR/UV double resonance spectroscopy in combination with REMPI spectroscopy is a very valuable tool to gain new information on intra- and intermolecular interactions. In the case of 1-methylthymine we were able to give a deeper insight to the properties of the dark intermediate state, which can be very helpful to unravel it's nature. For the clusters formed by CH...π and CH...N interactions, we could help to build up some very basic features of cluster and nano-crystal formation. These more profound results can help to gain a deeper understanding of the processes occurring in crystal formation.

Bibliography

[1] HERMAN, R.; Hepp M.; Hurtmans D.: High resolution Fourier transform spectroscopy of jet-cooled molecules. In: *Int. Reviews in Physical Chemistry* 19 (2000), S. 277–325

[2] DEMTRÖDER: *Laserspektroskopie*. Springer, 2007

[3] HECHT, J.: *The laser guidebook*. 2. McGraw Hill, 1992

[4] STRUVE, Dr. B.: *Laser, Grundlagen, Komponenten, Technik*. 1. Technik Berlin, 2001

[5] GLASER, Wolfgang: *Photonik für Ingenieure*. 1. Verlag Technik Berlin, 1997

[6] HECHT, E.: *Optics*. 4. Addison Wesley, 2002

[7] LIPSON ; TANNHAUSER: *Optical physics*. Cambridge University Press, 2001

[8] MAX BORN, Avadh Behari B.: *Principles of optics: electromagnetic theory of propagation, interference and diffraction of light*. Cambridge University Press, 1999

[9] SCHREIER, W. J. ; SCHRADER, T. E. ; KOLLER, F. O. ; GILCH, P. ; CRESPO-HERNÁNDEZ, C. E. ; SWAMINATHAN, V. N. ; CARELL, T. ; ZINTH, W. ; KOHLER, B.: 315 (2007), 625 S

[10] MANTIONE, M. J. ; PULLMAN, B.: 91 (1964), 387 S

[11] RAHN, R. O. ; PATRICK, M. H.: Photochemistry and photobiology nucleic acid. In: WANG, S. Y. (Hrsg.): *Biologie* Ed. II. Academic Press, Inc., New York, 1976, S. 97–145

[12] BECKER, R. ; KOGAN, G.: 31 (1980), 5 S

[13] CALLIS, P. R.: 34 (1983), 329 S

[14] CLARK, L. B. ; PESCHEL, G. G. ; I. TINOCO, Jr.: 69 (1965), 3615 S

[15] FUJII, M. ; TAMURA, T. ; MIKAMI, N. ; ITO, M.: Electronic spectra of uracil in a supersonic jet. 126 (1986), Nr. 6, S. 583–587

[16] TSUCHIYA, Y. ; TAMURA, T. ; FUJII, M. ; ITO, M.: 92 (1988), 1760 S

[17] VOET, D. R. ; GRATZER, W. B. ; COX, R. A. ; DOTY, P.: 1 (1963), 193 S

[18] BRADY, B. B. ; PETEANU, L. A. ; LEVY, D. H.: 147 (1988), 538 S

[19] LORENZON, J. ; FÜLSCHER, M. P. ; ROOS, B. O.: 117 (1995), 9265 S

[20] MARIAN, C. M. ; SCHNEIDER, F. ; KLEINSCHMIDT, M. ; TATCHEN, J.: 20 (2002), 357 S

[21] PETKE, J. D. ; MAGGIORA, G. M. ; CHRISTOFFERSEN, R. E.: 96 (1992), 6992 S

[22] CRESPO-HERNÁNDEZ, C. E. ; COHEN, B. ; KOHLER, P. M.: 104 (2004), 1977 S

[23] HARE, P. M. ; CRESPO-HERNÁNDEZ, C. E. ; KOHLER, B.: In: *PNAS* 104 (2007), 435 S

[24] SONG, Q. ; LIN, W. ; YAO, S. ; LIN, N.: 114 (1998), 181 S

[25] NIKOGOSYAN, D. N. ; LETOKHOV, V. S.: 8 (1983), 1–72 S

[26] GUERON, M. ; EISINGER, J. ; LAMOLA, A. A.: In: TS'O, Paul O. P. (Hrsg.):

Basic Principles in Nucleic Acis Chemistry. Academic Press, Inc., New York, 1974, S. 311–398

[27] SALET, C. ; BENSASSON, R.: 22 (1975), 231 S

[28] SALET, C. ; BENSASSON, R. V. ; BECKER, R. S.: 30 (1975), 325 S

[29] WOOD, P. D. ; REDMOND, R. W.: 118 (1996), 4256 S

[30] MARGUET, S. ; MARKOVITSI, D.: 127 (2005), 5780 S

[31] VIANT, Mark R. ; FELLERS, Raymond S. ; MCLAUGHLIN, Ryan P. ; SAYKALLY, Richard J.: Infrared-Laser Spectroscopy of Uracil in a Pulsed Slit Jet. 103 (1995), Nr. 21, S. 9502–9505

[32] BROWN, R. D. ; GODFREY, P. D. ; MCNAUGHTON, D. ; PIERLOT, A. P.: 110 (1988), 2329 S

[33] HE, Y. ; WU, C. ; KONG, W.: 108 (2004), 943 S

[34] YOSHIKAWA, A. ; MATSIKA, S.: Excited electronic states and photophysics of uracil-water complexes. 347 (2008), Nr. 1–3, S. 393–404

[35] JANZEN, C. ; SPANGENBERG, D. ; ROTH, W. ; KLEINERMANNS, K.: Structure and vibrations of phenol·$(H2O)_{7,8}$ studied by infrared-ultraviolet and ultraviolet-ultraviolet double-resonance spectroscopy and ab initio theory. 110 (1999), Nr. 20, S. 9898–9907

[36] KLEINERMANNS, K. ; JANZEN, Ch. ; SPANGENBERG, D. ; GERHARDS, M.: Infrared Spectroscopy of Resonantly Ionized (Pehnol)$(H_2O)_n^+$. 103 (1999), Nr. 27, S. 5232–5239

[37] PERUN, S. ; SOBOLEWSKI, A. ; DOMCKE, W.: Conical Intersections in Thymine. 110 (2006), S. 13238

[38] MATSIKA, S.: Three-state conical intersections in nucleic acids bases. 109

(2005), S. 7538

[39] MERCHÁN, M. ; GONZÁLEZ-LUQUE, R. ; CLIMENT, T. ; SERRANO-ANDRÉS, L. ; RODRÍGUEZ, E. ; REGUERO, M. ; PELÁEZ, D.: Unified Model for the Ultrafast Decay of Pyrimidine Nucleobases. 110 (2006), S. 26471

[40] PECOURT, Jean-Marc L. ; PEON, Jorge ; KOHLER, Bern: DNA Excited-State Dynamics: Ultrafast Internal Conversion and Vibrational Cooling in a Series of Nucleosides. 123 (2001), Nr. 42, S. 10370–10378

[41] AHLRICHS, R. ; BÄR, M. ; HÄSER, M. ; HORN, H. ; KÖLMEL, C.: 162 (1989), 165–169 S

[42] WERNER, H.-J. ; KNOWLES, P. J. ; LINDH, R. ; MANBY, F. R. ; SCHÜTZ, M. ; CELANI, P. ; KORONA, T. ; RAUHUT, G. ; AMOS, R. D. ; BERNHARDSSON, A. ; BERNING, A. ; COOPER, D. L. ; DEEGAN, M. J. O. ; DOBBYN, A. J. ; ECKER, F. ; HAMPEL, C. ; HETZER, G. ; LLOYD, A. W. ; MCNICHOLAS, S. J. ; MAYER, W. ; MURA, M. E. ; NICKLASS, A. ; PALMIERI, P. ; PITZER, R. ; SCHUMANN, U. ; STOLL, H. ; STONE, A. J. ; TARRONI, R. ; THORSTEINSSON, T.: "MOLPRO *version 2006.1, A package of ab initio programs.*". http://www.molpro.net, 2006

[43] HÄTTIG, C. ; WEIGEND, F.: CC2 Excitation Energy Calculations on Large Molecules Using the Resolution of the Identity Approximation. 113 (2000), S. 5154

[44] HÄTTIG, C. ; KÖHN, A.: Transition Moments and Excited-State First-Order Properties in the Coupled-Cluster Model CC2 Using the Resolution-of-the-Identity Approximation. 117 (2002), S. 6939

[45] HÄTTIG, C.: Geometry Optimizations with the Coupled-Cluster Model CC2 Using the Resolution-of-the-Identity Approximation. 118 (2003), S. 7751

[46] FLEIG, T. ; KNECHT, S. ; HÄTTIG, C.: Quantum-Chemical Investigation of the Structures and Electronic Spectra of the Nucleic Acid Bases at the Coupled Cluster CC2 Level. 111 (2007), S. 5482

[47] WERNER, H.-J. ; KNOWLES, P. J.: 82 (1985), 5053 S

[48] KNOWLES, P. J. ; WERNER, H.-J.: An Efficient Second-order MC SCF Method for Long Configuration Expansions. 115 (1985), S. 259

[49] CELANI, P. ; WERNER, H.-J.: Multireference Perturbation Theory for Large Restricted and Selected Active Space Reference Wave Functions. 112 (2000), S. 5546

[50] HEHRE, W.J. ; DITCHFIELD, R. ; POPLE, J.A.: Self-Consistent Molecular Orbital Methods. XII. Further Extensions of Gaussian–Type Basis Sets for Use in Molecular Orbital Studies of Organic Molecules. 56 (1972), S. 2257

[51] DESIRAJU, G. R. ; STEINER, T.: *The Weak Hydrogen Bond*. Oxford University Press, New York, 1999

[52] SCHEINER, S.: *Hydrogen Bonding: A Theoretical Perspective*. Oxford University Press, New York, 1997

[53] V.R. THALLADI, R.Boese: In: *New J. Chem.* 24 (2000), 463 S

[54] WHEATLEY, P. J.: In: *Acta Cryst.* 10 (1957), 182 S

[55] WITH, D. F. d.: In: *Acta Cryst. B* 32 (1976), 3178 S

[56] LAW, K. S. ; SCHAUER, M. ; BERNSTEIN, E. R.: Dimers of aromatic molecule: $(\mathrm{Benzene})_2$, $(\mathrm{toluene})_2$, and benzene-toluene. 81 (1984), Nr. 11, S. 4871–4882

[57] BÖRNSEN, K. O. ; SELZLE, H. L. ; SCHLAG, E. W.: Spectra of isotopically mixed benzene dimers: Details on the interaction in the vdW bond. 85 (1986), Nr. 4, S. 1726–1732

[58] HENSON, B. F. ; HARTLAND, G. V. ; VENTURO, V. A. ; FELKER, P. M.: 97 (1992), 2189 S

[59] SCHERZER, W. ; KRÄTZSCHMAR, O. ; SELZLE, H. L. ; SCHLAG, E. W.: 47 (1992), 1248 S

[60] VENTURO, V. A. ; FELKER, P. M.: 99 (1993), 748 S

[61] SCHMIED, Roman ; ÇARÇABAL, Pierre ; DOKTER, Adriaan M. ; LONIJ, Vincent P. A. ; LEHMANN, Kevin K. ; SCOLES, Giacinto: UV spectra of benzene isotopomers and dimers in helium nanodroplets. 121 (2004), Nr. 6, S. 2701–2710

[62] . WANNA, E. R. B.: In: *J, J. Chem. Phys.* 85 (1986), 777 S

[63] J. WANNA, E. R. B.:

[1] , 85 (1986) 3243. In: *J. Chem. Phys.* 85 (1986), S. 3243

[64] B. K. MISHRA, N. S.: In: *J. Theo. Comp. Chem.* 5 (2006), 609 S

[65] HÄBER, Thomas ; SEEFELD, Kai ; KLEINERMANNS, Karl: Mid and Near-Infrared spectra of conformers of H-Pro-Trp-OH. 111 (2007), Nr. 16, S. 3038–3046. `http://dx.doi.org/10.1021/jp070571e`. – DOI 10.1021/jp070571e

[66] HÜNIG, I. ; KLEINERMANNS, K.: Conformers of the peptides glycine-tryptophan, tryptophan-glycine and tryptophan-glycine-glycine as revealed by double resonance laser spectroscopy. 6 (2004), Nr. 10, S. 2650–2658

[67] JANZEN, C. ; SPANGENBERG, D. ; ROTH, W. ; KLEINERMANNS, K.: Structure and vibrations of phenol·(H2O)$_{7,8}$ studied by infrared-ultraviolet and ultraviolet-ultraviolet double-resonance spectroscopy and ab initio theory. 110 (1999), Nr. 20, S. 9898–9907

[68] GERHARDS, Markus: High energy and narrow bandwidth mid IR nanosecond

laser system. 241 (2004), S. 493–497

[69] F. WEIGEND, M. H.: In: *Theo. Chem. Acc.* 97 (1997), 331 S

[70] F. WEIGEND, H. P. R. A.: In: *Chem. Phys. Lett.* 294 (1998), 143 S

[71] GRIMME, S.: In: *J. Comput. Chem.* 27 (2006), 1787 S

[72] GRIMME, S.: In: *J. Comput. Chem.* 25 (2004), 1463 S

[73] J. F. ARENAS, J. C. I. M.: In: *J. Chem. Soc., Faraday Trans. 2* 81 (1985), 405 S

[74] J. E. PARKIN, K. K. I.: In: *J. Mol. Spectrosc.* 15 (1965), 407 S

[75] K. B. HEWETT, C. L. L. A. P.: In: *J. Chem. Phys.* 100 (1994), 4077 S

[76] R. H. PAGE, Y. T. L.: In: *J. Chem. Phys.* 88 (1988), 4621 S

[77] M. PIACENZA, S. G.: In: *Chem. Phys. Chem.* 6 (2005), 1554 S

[78] P. JUREČKA, J. ˙ H.: In: *Phys. Chem. Chem. Phys.* 8 (2006), 1985 S

[79] C. MORGADO, I. H. S.: In: *Phys. Chem. Chem. Phys.* 9 (2007), 448 S

[80] T. SCHWABE, S. G.: In: *Phys. Chem. Chem. Phys.* 9 (2007), 3397 S

[81] P. HOBZA, E. W. S.: In: *J. Phys. Chem.* 100 (1996), 18790 S

[82] T. N. WASSERMANN, M. A. D. L.: In: *J. Chem. Phys.* 127 (2007), 234309 S

[83] BOESE, R. ; CLARK, T. ; GAVEZZOTTI, A.: In: *Helvetica Chimica Acta* 86 (2003), 1085–1100 S

[84] R. BOESE, W. E. Billups L. R. N.: In: *Angew. Chemie Int. Ed* 42 (2003), 1961–1963 S

[85] KODAMA, Y. ; NISHIHATA, K. ; NISHIO, M. ; IITAKA, Y.: 2 (1979), 1490 S

[86] NISHIO, M. ; HIROTA, M. ; UMEZAWA, Y.: *The CH/π interaction: Evidence,*

Nature and Consequences. Wiley-VCH, New York, 1998

[87] NISHIO, M.: In: *Cryst. Eng. Comm.* 6 (2004), 130 S

[88] BRANDL, M. ; WEISS, M. S. ; JABS, A. ; SÜHNEL, J. ; HILGENFELD, R.: 307 (2001), 357 S

[89] FUJII, A. ; MORITA, S. ; MIYAZAKI, M. ; EBATA, T. ; MIKAMI, N.: 108 (2004), 2652–2658 S

[90] SAMPSON, R. K. ; BELLM, S. M. ; GASCOOKE, J. R. ; LAWRENCE, W. D.: 372 (2003), 307–377 S

[91] BRAND, J. C. D. ; EGLINTON, G. ; TYRRELL, J.: (1965), 5914–5919 S. http://dx.doi.org/10.1039/JR9650005914. – DOI 10.1039/JR9650005914

[92] SUNDARARAJAN, K. ; VISWANATHAN, K. S. ; KULKARNI, A. D. ; GADRE, S. R.: 613 (2002), 209–222 S

[93] NAKAGAWA, N. ; FUJIWARA, S.: 33 (1960), 1634–1634 S

[94] CARRASQUILLO, E. ; ZWIER, T. S. ; LEVY, D. H.: 83 (1985), 4990–4999 S

[95] SHELLEY, M. Y. ; DAI, H.-L. ; TROXLER, T.: 110 (1999), 9081–1990 S

[96] SHIBASAKI, K. ; FUJII, A. ; MIKAMI, N. ; TSUZUKI, S.: 111 (2007), 753–785 S

[97] GOTCH, A. J. ; ZWIER, T. S.: 96 (1992), 3388–3401 S

[98] GARRETT, A. W. ; ZWIER, T. S.: 96 (1992), 3402–3410 S

[99] TSUZUKI, S. ; HONDA, K. ; UCHIMARU, T. ; MIKAMI, M. ; TANABE, K.: 122 (2000), 3746–3753 S

[100] NOVOA, J. J. ; MOTA, F.: 318 (2000), 345–354 S

[101] OKRUSS, M. ; MÜLLER, R. ; HESE, A.: High-Resolution UV Laser Spec-

troscopy of Jet-Cooled Benzene Molecules: Complete Rotational Analysis of the $S_1 \leftarrow S_0 6_0^1 (l = \pm 1)$ Band. 193 (1999), Nr. 2, S. 293–305

[102] Busker, Matthias ; Häber, Thomas ; Nispel, Michael ; Kleinermanns, Karl: Isomer-Selective Vibrational Spectroscopy of Benzene-Acetylene Aggregates: Comparison with the Structure of the Benzene-Acetylene Cocrystal. In: *Angew. Chemie Int. Ed* 47 (2008), S. 10094–10097. http://dx.doi.org/10.1002/anie.200802118. – DOI 10.1002/anie.200802118

[103] Montero, L. A.: *GRANADA and Q programs for PC computers ed.* 1996

[104] Stewart, James J. P.: Optimization of Parameters for Semiempirical Methods V: Modification of NDDO Approximations and Application to 70 Elements. In: *J. Mol. Modeling* 13 (2007), S. 1173–1213

[105] Stewart, James J. P.: *MOPAC 2007, Stewart Computational Chemistry, Colorado Springs, CO, USA, http://OpenMOPAC.net.* 2007

Used Abbreviations and Symbols

Abbreviation/ Symbol	Meaning
1-MT	1-Methylthymine
1-MU	1-Methyluracil
A	Acetylene
B	Benzene
BBO	β-Bariumborate
c	Speed of light
CI	Conical intersection
C_p	Heat capacity at constant pressure
C_V	Heat capacity at constant volume
CPD	Cyclobutane pyrimidine dimer
DNA	Deoxyribonucleic acid
DFG	Difference frequency generation
DFM	Difference frequency mixing
DFT	Density functionmal theory
E	Energy
Excimer	Excited state dimer
f	Molecular energetic degrees of freedom
GPIB	Computer interface (IEEE 488.2)
h	Planck's constant

Abbreviation/ Symbol (cont.)	Meaning (cont.)
h	Mass dependant enthalpy
HV	High voltage
IC	Internal conversion
IP	Ionization potential
IR	Infrared light
ISC	Inter system crossing
IVR	Intramolecular vibrational redistribution
k	Boltzmann's constant
$K_{n,0}$	Knudsen number
KDP	Potassiumdihydrogenphosphate
KTA	Potassiumtitanylarsenate
KTP	Potassiumtriphosphate
LIF	Laser induced fluorescence
M	Mach's number
M	Molar mass
MCP	Multi channel plate
NMR	Nuclear magnetic resonance
Ns-TAS	Nanosecond transient absorption spectroscopy
OPA	Optical parametric amplifier
OPO	Optical parametric oscillator
PC	Personal computer
Pd	Pyridine
P_r	Residual pressure
Pz	Pyrazine
R	Ideal gas constant
R2PI	Resonant two photon ionization
REMPI	Resonant multi photon ionization

Abbreviation/ Symbol (cont.)	Meaning (cont.)
RIDIR	Resonant ion depletion infrared spectroscopy
R_s	Specific gas constant
SFB	Sonderforschungsbereich
SHB	Spectral Hole Burning
SHG	Second Harmonic Generation
T	Thymine
T	Temperature
t	time
TMP	Thymidine monophosphate
TOF-MS	Time of flight mass spectrometer
TTL	Transistor transistor logic
u	Maximum jet velocity
U	Uracile
U	Voltage
UMP	Uracile monophosphate
UV	Ultraviolet light
VIS	Visible light

List of Figures

Acknowledgements

Diese Arbeit wäre sicherlich nicht ohne den Beistand und die Hilfe vieler Menschen entstanden, denen ich hier danken möchte.

Meinem Doktorvater Karl Kleinermanns möchte ich für die Chance als Quereinsteiger in seinem Institut promovieren zu dürfen danken. Weiterhin bin ich sehr dankbar für die vorgegebenen Themen, die in Kombination mit zahllosen Diskussionen und neuen Denkanstößen schließlich zu der hier vorliegenden Arbeit und den erfolgreichen Publikationen geführt haben.

Weiterhin danke ich Michael Schmitt für die Übernahme des Zweitgutachtens, die Begleitung bei meinen Promotionseignungsprüfungen und den vielen kollegialen Hilfestellungen.

Thomas Häber danke ich für die sehr entspannte und überaus angenehme Zusammenarbeit, während der ich im Büro ”‘en passant”’ unglaublich viel gelernt habe. Nicht zu vergessen, dass erst Thomas durch seine Rechnungen die ausführliche Diskussion der vorliegenden Daten ermöglicht hat.

Klaus Kelbert und Dieter Marx danke ich für ihre Hilfe bei elektronischen, mechanischen und allen anderen Problemen im Labor und die angenehmen Gespräche.

Allen Mitarbeitern der mechanischen Werkstatt danke ich sehr für Ihr Bemühen, meine Vorstellungen in die Realität umzusetzen.

Ausserdem danke ich meinen Mitstreitern der Physikalischen Chemie I (Thomas

Häber, Gernot Engler, Kai Seefeld, Katharina Hunger, Michaela Braun, Ke Feng, Thi Bao Chau Vu, Christian Brand, Nathalie Fröse, Olivia Noga, Daniel Ogermann, Thorsten Wilke und Lars Biemann) für das sehr lustige und eigentlich schon freundschaftliche Verhältnis in der Arbeitsgruppe und die unedlich vielen Kleinigkeiten, die einem den Alltag vereinfachen.

Ganz besonders danke ich meinen Eltern Helene und Göke Busker für die andauernde Hilfe, Ünterstützung und Liebe, die sie mir während meines 22 jährigen Schul- und Bildungsweges gegeben haben. Weiterhin danke ich meinem Bruder Stephan und meiner gesamten Familie für ihre Unterstützung. Ohne euch alle wäre diese Arbeit sicher nicht möglich gewesen.

Zu guter letzt danke ich meiner Freundin Carola, die mich schon so lange begleitet und mir mit ihrer unverblümten Meinung, ihrer starken Unterstützung und ihrer Liebe hoffentlich noch viele Jahrzehnte den Weg durch das Leben weist.

Danke Euch allen!

Printed by Books on Demand GmbH, Norderstedt / Germany